CW01024466

DRUMMED OUT!

THE SACKING OF PETE BEST

SPENCER LEIGH

First Published in 1998 by
NORTHDOWN PUBLISHING LIMITED
PO Box 49, Bordon
Hants GU35 0AF

British Library Cataloguing-in-Publication Data
A catalogue for this book is available from the British Library

ISBN 1900711 04 4

Designed by Ian Welch

Cover by Jimmy Egerton

PICTURE CREDITS
Photographs in this book have been reproduced courtesy of:
Pictorial Press, Johnny Guitar, Steve Hale and Spencer Leigh.

CONTENTS

Drummed Out – The Sacking Of Pete Best *is dedicated to my wife,
Anne, who says that I suffer from Parkinson's disease –
Michael Parkinson's, that is – an overwhelming desire to
interview everyone I meet.*

MAY I INTRODUCE TO YOU

"Think what we would have missed if we had never heard the Beatles."
Queen Elizabeth II, celebrating her golden wedding.

The Beatles' career has been documented in thousands of books, articles and broadcasts, not to mention the Internet. A lifetime would be too short to digest it all, but I am certain that no-one has seriously attempted to explain the sacking of Pete Best.

So much is known about the Beatles and yet so little is known about the background to Pete Best's dismissal:

* Why did the other Beatles sack Pete Best?
* What role did each individual Beatle play in that decision?
* Did Brian Epstein encourage them or try to dissuade them?
* Why was Pete Best sacked in such an insensitive manner?
* Was Ringo Starr the obvious replacement?

Only four people know with certainty what went on – John Lennon, Paul McCartney, George Harrison and Brian Epstein. Apart from a few cryptic comments, the normally garrulous Lennon died with his story. Ditto Brian Epstein. There is much that Paul McCartney and George Harrison can say about this issue, but it seems unlikely that they will do at this late stage. Possibly they will confirm or deny this book's conclusions, but don't hold your breath. Both are skilful at dodging reporters' questions and, in court in May 1998, George said he used meditation to help forget the past.

The Beatle literature is also silent on this subject. The contemporary newspaper, *Mersey Beat*, was subjected to spin-doctoring by Brian Epstein. The prolific Beatle biographer, Geoffrey Giuliano, skirts the subject, while Alan Clayson gives one explanation in his biography of George Harrison and another in his biography of Ringo Starr. Interviewers have missed their opportunities, although, to be fair, the issue may not be high in their list of questions. Jann Wenner for *Rolling Stone* (1971), David Sheff for *Playboy* (1980) and Andy Peebles for the BBC (1980) spent hours interviewing John Lennon, and all failed to raise the issue. I can't criticise them as I'm as neglectful myself. I had 22 minutes of Paul McCartney's time for BBC Radio Merseyside and never even mentioned their hapless drummer. Add to this, that Paul is so damn nice. I suspect that even Jeremy Paxman would submit to his charm and not pursue his interrogation.

The publication of Paul McCartney and Barry Miles' *Many Years From Now* in 1997 prompted this book. Paul is frank, very frank, about his sexual exploits and his drug-taking but, despite 600 pages, Pete Best only merits one line in the index – to be accurate, only half a line as he is bracketed with his mother Mona, usually known as Mo. We learn that the Beatles went to Mo's club, the Casbah, that Pete joined the group, and that Pete left two years later. That's it, and yet there are whole pages about their former bass guitarist, Stuart Sutcliffe, who, musically speaking, was a passenger. To be fair, the publisher may have cut the text or Paul may be saving the Pete Best saga for a second volume or another project but I doubt it.

So I decided to write *Drummed Out*. I have spoken to numerous musicians about Pete Best and I have looked at all the major reference books and been through all the interviews I have kept over the years. I was hoping that this wealth of material would suggest some explanations and I have followed up my leads with further interviews. I have enjoyed being Inspector Morse. I hope that you agree with my conclusions, but whatever, you now have the information and can interpret it for yourself.

I have known rock writer Michael Heatley for several years, and our friendship blossomed when we wrote *Behind The Song* for Blandford Books in 1997. Michael has his own company, Northdown Publishing and I am proud to be published alongside the first-rate and very distinctive books on Dr Feelgood, Man and Ralph McTell.

Unless identified otherwise, all the quotes come from interviews I have conducted for BBC Radio Merseyside, or for the annual Merseybeatle Convention or for this book. My thanks to BBC Radio Merseyside for the use of my interview material and also to David Horn at the Institute of Popular Music at the University of Liverpool for encouragement and for putting me in contact with an authority on drumming techniques, Garry Tamlyn, the Head of Contemporary Music at Queensland Conservatorium, Australia. My thanks also to Trevor Cajiao, Andrew Doble, Peter Doggett, Bob Groom, Neil Hiley, Ian Kennedy, Bernd Matheja, Bill Morrison, Mick O'Toole, Denis Reed, Daniel van der Slik, René van Haarlem and Granville Wolstenholme. I have also used an interview with Pete and Roag Best for the 17th International Dutch Beatles Convention, Amsterdam in 1996.

Because Pete Best and his brother Roag are involved in various projects, including a film script, they felt unable to help with *Drummed Out*, but I have drawn upon personal interviews over the years as well as Pete's autobiographies. There can't be many people who have written their life story twice. How many more does he plan?

Whatever Pete Best's limitations, the Beatles treated him shittily. His pride took a tremendous knock and yet he has acted without bitterness or rancour over the years. If he harbours a grudge, he keeps it well hidden. Large portions of this book may be as new to Pete as they were to me – after all, when you're kicked out of a band, you're not party to the discussions which explain why you're being sacked.

Spencer Leigh

PAPERBACK WRITERS

The following books have been consulted in researching this book:

Best, Pete & Patrick Doncaster	*Beatle! The Pete Best Story*	(Plexus, 1985)
Best, Pete & Bill Harry	*The Best Years Of The Beatles*	(Headline, 1996)
Brown, Peter & Steven Gaines	*The Love You Make*	(Macmillan, 1983)
Clayson, Alan	*The Quiet One – A Life Of George Harrison*	(Sidgwick and Jackson, 1990)
Clayson, Alan	*Ringo Starr – Straight Man Or Joker?*	(Sidgwick and Jackson, 1991)
Coleman, Ray	*John Winston Lennon*	(Sidgwick and Jackson, 1984)
Coleman, Ray	*Brian Epstein* *The Man Who Made The Beatles*	(Viking, 1989)
Davies, Hunter	*The Beatles – The Authorised Biography*	(Heinemann, 1968)
Du Noyer, Paul	*We All Shine On*	(Carlton, 1997)
Foster, Mo	*Seventeen Watts?*	(Sanctuary, 1997)
Frame, Pete	*The Beatles And Some Other Guys*	(Omnibus Press, 1997)
Goldman, Albert	*The Lives Of John Lennon*	(Bantam, 1988)
Gottfridsson, Hans Olof	*The Beatles – From Cavern To Star-Club*	(Premium Publishing, 1997)
Giuliano, Geoffrey	*Tomorrow Never Knows* *30 Years Of Beatles Music And Memorabilia*	(Paper Tiger, 1991)
Harry, Bill	*The Ultimate Beatles Encyclopedia*	(Virgin, 1992)
Harry, Bill	*The Encyclopedia Of Beatles People*	(Blandford, 1997)
Hertsgaard, Mark	*A Day In The Life* *The Music And Artistry Of The Beatles*	(Macmillan, 1995)
Howlett, Kevin	*The Beatles At The Beeb*	(BBC, 1982)
Leigh, Spencer & Pete Frame	*Let's Go Down The Cavern*	(Vermilion, 1984)
Leigh, Spencer	*Speaking Words Of Wisdom*	(Cavern City, 1991)
Lewisohn, Mark	*The Beatles Live!*	(Pavilion, 1986)
Lewisohn, Mark	*The Complete Beatles Chronicle*	(Pyramid, 1992)
McCartney, Paul & Barry Miles	*Many Years From Now*	(Secker and Warburg, 1997)
MacDonald, Ian	*Revolution In The Head* *The Beatles' Records And The Sixties*	(Fourth Estate, 1994)
Marsden, Gerry & Ray Coleman	*I'll Never Walk Alone*	(Bloomsbury, 1993)
Peebles, Andy	*The Lennon Tapes*	(BBC, 1981)
Norman, Philip	*Shout! The True Story Of The Beatles*	(Hamish Hamilton, 1981)
Sheff, David	*The Playboy Interviews With John Lennon And Yoko Ono*	(Playboy Press, 1981)
Shotton, Pete & Nicholas Schaffner	*John Lennon In My Life*	Coronet (London, 1984)
Taylor, Alistair & Martin Roberts	*Yesterday – The Beatles Remembered*	(Sidgwick and Jackson, 1988)
Thompson, Phil	*The Best Of Cellars*	(Bluecoat Press, 1994)
Wenner, Jann	*Lennon Remembers* *The Rolling Stone Interviews*	(Straight Arrow, 1971)
Williams, Allan & William Marshall	*The Man Who Gave The Beatles Away*	(Elm Tree, 1975)

LET THERE BE DRUMS

'Ah, the Beatles. There were five of them if I remember.'
Dirk Bogarde being interviewed on BBC-TV, 1983

I

Two childhood reminiscences.

First verse: from Hunter Davies' authorised biography of the Beatles, published in 1968. Paul McCartney is talking about his relationship with his parents. 'I was pretty sneaky,' says Paul, 'If I ever got bashed for being bad, I used to go into their bedroom when they were out and rip the lace curtains at the bottom, just a little bit, then I'd think, that's got them.'

To be followed by 'The Frog Chorus' from Paul McCartney's authorised biography, *Many Years From Now*, published in 1997. Paul McCartney is going for the countryside vote. 'All my mates killed frogs anyway. They used to blow them up by sticking a straw up their ass. That was the way to kill a frog. I didn't fancy that, I thought that was a little bit pervy. I thought a straightforward killing with a bash, hold the legs and just smash 'em on the head.'

A maxim from William Wordsworth:
The child is the father to the man.

II

In July 1997, the five surviving Quarry Men reformed for a garden fête at St Peter's Church in Woolton. The opening number, quite appropriately, was 'Lost John', and it was followed by 'Midnight Special', 'Pick A Bale Of Cotton' and some rock'n'roll standards. The 50-somethings were recreating the skiffle music of their youth in order to recapture the moment when John Lennon met Paul McCartney 40 years earlier. It was a celebration for all who attended except, perhaps, the Quarry Men themselves. All but one had been sacked by John Lennon – the fifth, Len Garry, left because he contracted TB. But, TB or not TB, John Lennon was the only original Quarry Man remaining by 1959, some 18 months later.

On a film clip in the *Anthology* series, John Lennon stated, 'I was the singer and I was the leader and I made the decision to have Paul in the group. Was it better to have a guy who was better than the people I had in, or not? The decision was to let Paul in to make the group stronger.'

The Quarry Men's banjo player, Rod Davis, recalls, 'I had bought the banjo from my uncle and if he'd sold me his guitar, I might have been a decent enough guitarist to keep McCartney out of the band. I might have learnt guitar chords, I might not, and that was the big limitation really. McCartney could play the guitar like a guitar and we couldn't, and let's face it, a banjo doesn't look good in a rock'n'roll group. I only met Paul on one other occasion after the Woolton fête and it was at Auntie Mimi's a week or two later. He dropped in to hear us practising. From my point of view, I was the person he was replacing – it's like Pete Best – you're the guy who doesn't know. Some things had gone on that I was unaware of.'

John's best friend, Pete Shotton, played washboard. 'We were doing a gig in Rosebery Street, it was Colin's auntie's do. We were performing on the back of a lorry and John and I went inside the house and we were sitting on the floor. We were having a few beers, John has a reputation for drinking but he hadn't been drinking when Paul first met him. We were drinking a few beers and we were getting pissed because it only takes a couple at that age. I said to John, "This is not really my scene, I'm embarrassed to be up there." John picked up my washboard and smashed it over my head and said, "That solves that then, Pete." The decision had already been made that he was going to develop a serious band and that couldn't, by definition, include me. It didn't hurt and we fell about laughing – me with relief and John with relief as well. He had resolved a situation which was very tricky for him.'

Eric Griffiths hung on for six months, trying to play the guitar competently. 'Paul had mentioned George Harrison in the context of him being a particularly good guitarist. I don't think I ever played any shows with him but I certainly practised with him. He was a good guitarist, better than me, and shortly after that I left the group. Paul and John asked me to go on to the bass and that meant buying a bass guitar and an amplifier which I wasn't prepared to do as the group wasn't going anywhere and it was an investment on my mother's part that I wasn't prepared to follow through. So I went and George was in. I was sorry to go and they then had to decide who was going to play the bass out of the three of them. I never played a guitar again. I joined the Merchant Navy and my life went in a different direction.'

Drummer Colin Hanton: 'John, Paul and Eric Griffiths were the three guitarists and I was on drums. Eventually we were playing at the Morgue in Old Swan, which was run by Rory Storm, and it was a real dump. It was a condemned building and the whole thing was illegal and never advertised. It was a large terraced house and the front room was being used as a tiny dance-hall with a stage. There was a long corridor to other rooms which were the dressing-rooms, and that is where we met George Harrison. Somebody asked him to play something. I thought it was "Guitar Boogie" but everyone else reckons it was "Raunchy". A few days later Ivan Vaughan told me that Paul and John wanted George to join the group, but Eric would have to go as they didn't want four guitarists. I was living on borrowed time as they were running the group. I got fed up in the end. I had carted my drums around on a bus for two years as none of us had cars. There were a lot of talent contests where we came second – we were always the bridesmaid and I'd had enough.'

Friendships didn't count. John Lennon would replace musicians if he thought it would improve the group. In the future, if the occasion arose, he would be prepared to sack Pete Best.

III

An English army officer, John Best, married a Red Cross nurse of English parentage, Mona, when he was in India. Randolph Peter Best was born in Madras on 24 November 1941 and his brother, Rory, was born in 1944. The family arrived in Liverpool on Christmas Day 1945 and, two years later, settled in a large Victorian house in a middle-class suburban area – 8 Hayman's Green, West Derby. Johnny Best was a notable boxing promoter, staging fights at Liverpool Stadium by Randolph Turpin and Freddie Mills. Their marriage fell apart in the early 1950s, and Mona devoted herself to her sons and their friends, who called her 'Mo', a derivative of mother as well as Mona.

Peter won a scholarship to the Liverpool Collegiate in Shaw Street. He obtained five O levels and was half-heartedly thinking of becoming a teacher, an unlikely choice for someone depicted as 'silent' in all the Beatle books. Then rock'n'roll took its hold.

The jazz-based Cavern club had opened in 1957, but some coffee-bars were catering for rock'n'roll-minded teenagers. Mona Best turned the large cellars of her house into a rock'n'roll club. She named it, rather exotically, the Casbah, a reference to the 1938 film, *Algiers*, which starred Charles Boyer and which supposedly contains the line, 'Come with me to the Casbah', although this is never said.

Mona Best: 'The boys wanted a club for their friends but I thought it should be bigger than that. We worked all hours to get the cellar open and we had people coming from all places. Everyone wanted to join the Casbah. We ended up with 2,950 members, which was fantastic.' Fortunately, they all didn't attend at once but for a house in the suburbs, it was a remarkable achievement. It also illustrates how ill-advised the Cavern was to stick with jazz.

The Les Stewart Quartet was booked for the opening night, 29 August 1959. The Quarry Men had few bookings and George Harrison was one of the Quartet. However, a fierce argument led to Les Stewart disbanding the group, and George and their bass player, Ken Brown, teamed up with John and Paul to perform as the Quarry Men on the opening night. No drummer. John brought along his girlfriend, Cynthia Powell, whom he had met at the College of Art and whom he would eventually marry, and another art school friend, Stuart Sutcliffe.

The Quarry Men were booked for each subsequent Saturday night – the club was simply open as a coffee-bar with a jukebox for the rest of the week. On 10 October, Ken Brown had hurt his leg and was not fit to play with the Quarry Men. Their fee was £3 and rather than give £1 to John, Paul and George, Mo gave them each 15 shillings (75p) and said she was holding the balance for Ken. The three guitarists were so annoyed that they vowed never to play the Casbah again and sacked the ailing Mr Brown for hanging on to 15 bob.

Not to worry, as Ken had seen Pete Best bashing out the rhythm on chairs and tables. He suggested that they formed a group and so the Blackjacks was born. They once attracted 1,350 to the Casbah which shows you how popular they were, how big the house was and how cramped the conditions would be. Fire regulations were nowhere as stringent as today's. The Best's neighbours were a placid and tolerant bunch – partly because so few teenagers had cars and partly because their own children went to the Casbah. Mona Best: 'The club had a friendliness and a homeliness that was hard to equal. If there was any trouble at the Casbah, you didn't just take on the bouncers, you took on the club.'

Other groups to appear at the Casbah included Gerry and the Pacemakers and Rory Storm and the Hurricanes with their drummer, Ringo Starr. The three key members of the Quarry Men did not play the Casbah again until 17 December 1960, which marked their return from Hamburg with Pete Best as their drummer.

IV

Allan Williams reminds me of Arthur Daley in *Minder*. It may be unfair but the image is of his own making as the entrepreneur can be found in Liverpool pubs, dreaming up new ideas to make him millions. Somehow they never come off. His scheme to turn Spain into another Blackpool by manufacturing rock for tourists fell foul of the authorities, and people say he was on street corners going, 'Psst, want to buy a stick of rock.' Some old black leather trousers, allegedly worn by Paul McCartney, appeared in a Sotheby's catalogue with a certificate of authenticity from Allan Williams. It transpired that they belonged to Faron of Faron's Flamingos.

It was Allan Williams, a one-time plumber, who introduced bullfighting (with a real bull!) into Liverpool clubland. It was Allan Williams who booked the Beatles to back Janice, an over-endowed Manchester stripper, for a week in a dodgy Liverpool club. It was Allan Williams who planned to get the tapes of the Beatles in Hamburg released: he also planned to cut up the original tapes and sell one-inch strips in key-chain souvenirs. On the videotape of *Imagine...The Sixties*, Billy Butler refers to a club burning down, adding, 'And Allan Williams didn't own it.' When Allan asked Paul McCartney to sign his book about the Fab Four, *The Man Who Gave The Beatles Away*, Paul said, 'I've got to be careful here. Whatever I write is going to be quoted on the paperback.' He wrote, 'To Allan, Some parts of this book are partially true, Paul McCartney.'

Allan Williams and the former Cavern DJ Bob Wooler have often appeared at Beatle Conventions, even travelling to New York for a prestigious one. They stayed in the same hotel as Richard Nixon and several aides, and Allan and Bob had the larger bar bill, paid of course by the organisers. Nice to think that Brits can do some things better than the Yanks.

Every city presumably has one but Allan Williams is one of the most colourful characters around. I can't help liking the guy even though I'd never buy anything from him.

Footnote
I passed this text to Allan Williams before publication and asked him if there was anything he wished to amend, secretly hoping that he would add a few more hilarious misadventures. He told Bob Wooler that a few things were wrong – he was more like Delboy in Only Fools And Horses *than Arthur Daley – but he never gave me the amendments. Not that he would be bothered: when ghosting* The Man Who Gave The Beatles Away, *Bill Marshall of the* Daily Mirror *found that he had not enough material. He asked Allan if he could invent some stories about the Beatles to fill it out. Allan said yes and didn't even see them before publication. These stories have become part of Mythew Street and Allan Williams recounts them at Beatle conventions as though they really happened. Allan Williams is contemplating another semi-fictional book of Beatle reminiscences when he should be writing about his other business ventures. There's a wonderful book to be written about Liverpool's underworld in the late 1950s/early 1960s, a book that could do for Liverpool what Colin MacInnes'* Absolute Beginners *did for Soho.*

V

In 1958 Allan Williams, himself a trained singer, opened the Jacaranda coffee-bar in Liverpool's city centre – 23 Slater Street to be precise. It's still there and, in the basement, you can see the original murals by Stuart Sutcliffe and, possibly, John Lennon – well, guess who claims that? Don't all rush at once – they're nothing to write home about, just gloomy, sub-Picasso doodlings.

Early in 1960 Eddie Cochran and Gene Vincent were touring the UK for impresario Larry Parnes. The tour played the Liverpool Empire for a week and Allan Williams suggested that the two rockers might top a show at the Liverpool Stadium on their return in May. Cochran was killed on 17 April but Vincent was forced to honour his contract. The Stadium show included Gerry and the Pacemakers but not the Beatles, who were not considered good enough.

Parnes told Williams that he was looking for a backing group for Billy Fury, and the Beatles attended the auditions on 10 May with Johnny Hutchinson of Cass and the Cassanovas playing drums. As a result, the Beatles backed another Parnes protégé (and Scouser), Johnny Gentle, on a tour of Scotland. John, Paul, George and Stu (on bass) took Tommy Moore as their drummer.

Artist and blues singer Al Peters: 'Stu Sutcliffe was a bass player who was passing through and that's putting it kindly. He was in the band because he was a friend of John's. The first thing you do when you have a band is to have your friends in because you can relate to them easier than strangers. Also, John and Stu had artistic endeavours between them; they were kindred spirits and could talk about the same things. I saw the TV programme in 1997 about Paul McCartney's art work and if you compare that to the depth in Stu's work, you'll see the difference. Paul is a musician who is dabbling in art and if you reverse that, you have an artist who is playing a bit of bass to help his mates out. Paul is enjoying what he does, but how he arrives at his points on the canvas is totally different to way that Stuart arrived at his. Stuart had emotional and physical problems and that also shows in his work.'

Allan Williams: 'Nobody rated the Beatles at all early on, they were known as a rubbish group, and they had difficulty finding a drummer. Also, John was a very difficult person to get on with. Poor Tommy Moore was recommended by Cass of the Cassanovas. He was ten years older than John and stood no chance at all.'

As well as suffering John's vicious wit, Tommy Moore lost several teeth in a car accident. When John came to see him in hospital, he insisted he joined the band on stage in Fraserburgh that night. Back in Liverpool, the group played at the Jac and secured a booking at the Grosvenor Ballroom in Wallasey. When Moore didn't turn up, Allan Williams went to collect him from his home in Toxteth. 'You can piss off,' screamed his girlfriend. 'He's got a job on the night-shift at Garston Bottle Works.'

So ended the career of another Beatle drummer. Norman Chapman played a few gigs but was conscripted into the Army. However, my favourite has to be Rockin' Ronnie. John Lennon announced on stage that they were minus a drummer and could anyone help. Rockin' Ronnie stepped forward and although the huge Teddy Boy had never touched a set of drums before, the Beatles had to do the best they could – or suffer the consequences.

VI

Lord Woodbine, Allan Williams' front for his strip club, was forced to close it down. In May 1959, Allan also staged 'Stars Without Bras' compered by Larry Grayson at the Pavilion Theatre, Lodge Lane and ran into trouble as his nude girls were gyrating on stage instead of standing still. 'We sold the club and the show, and Woody and I went to Amsterdam on the dirty businessman's trip you could get for £10 return in an old Dakota. We moved on to Hamburg and I heard a German group singing rock'n'roll songs that they had learnt parrot-fashion. The Kaiserkeller agreed to take some Liverpool groups, but we couldn't agree a fee. Unknown to me, Bruno Koschmider went to London to find Rory Storm, Gerry and the Pacemakers, and the Beatles, and he was persuaded to take Tony Sheridan instead. Meanwhile, Derry Wilkie and the Seniors had been let down by Larry Parnes and blamed me for this. I took them to the Two I's in London and quite by chance, Bruno was there looking for some groups.' Koschmider, a Grade A thug and almost certainly ex-Gestapo, had had his testicles shot away in World War II. He must have looked out of place in that teenage coffee bar.

So, by the summer of 1960, Allan Williams was a booking agent for the Hamburg clubowner, Bruno Koschmider. The Seniors were performing with success at the Kaiserkeller and Koschmider wanted a second Liverpool group for the Indra. Williams needed a group whose members were not tied to day jobs and, having little choice, selected the Silver Beatles. There was a problem. The contract stipulated a drummer and the Beatles didn't have one. Sod the contract – if Bruno Koschmider wanted a drummer, you had to have one. In the *Anthology* series, Paul McCartney defended the drummerless Beatles: 'People would say, "Where's the drums?" and we would say, "The rhythm's in the guitars".' Such an argument wouldn't convince Koschmider. As an aside, it should be noted that the Hamburg club managers had wonderfully onomatopoeic names – Bruno Koschmider, Horst Fascher, Manfred Weissleder. In Liverpool, on a lighter note, there was George Blott and Sam Leach. You couldn't invent names like these.

The Silver Beatles consisted of four guitarists – John Lennon, Paul McCartney, George Harrison and Stu Sutcliffe – and they had to find a drummer. On 6 August a local booking had been cancelled and they gravitated to the Casbah. The Blackjacks were playing but because of college commitments, they were on the verge of breaking up. However, Pete Best, with his new kit, wanted to be a professional drummer.

A vintage interview with John Lennon was included in the *Anthology* series. 'People who owned drum-kits were few and far between because it was an expensive item and they were usually idiots, you know. We got Pete Best just because we needed a drummer for Hamburg.'

Harsh words – and how the film-makers must have been delighted to find that – but Allan Williams would agree, 'Pete Best wasn't a Beatle. When I got them the Hamburg job, they needed a drummer and they came up with Pete Best. He didn't fit in with their image or get on socially with them. I don't think he was sacked because of his drumming.'

VII

From 17 August to 3 October 1960, the Beatles played at the Indra on the Grosse Freiheit in Hamburg. Then they moved to the Kaiserkeller, playing alternate sets with Rory Storm and the Hurricanes until they were deported for alleged arson on 30 November. No Allan Williams jokes please. Actually, Paul and Pete had gone to their dilapidated sleeping quarters at Koschmider's mac brigade cinema, the Bambi Kino, to collect their belongings as they were moving to the new Top Ten club with Tony Sheridan. There was no light in the room so they attached contraceptives to some old tapestry on the walls and lit them. They gathered their belongings together while the contraceptives burnt out, leaving scorch marks on the wall. No harm done, certainly not in a dump like that, but Bruno did not want the Beatles playing at a rival club. He didn't want to start a gang war at that time: hence the deportation.

The Beatles had played 106 nights in Germany and their throats must have been like raw liver. It's hard to credit John Lennon's voice being on its last legs at a Parlophone recording session, and George Martin having to get 'Twist And Shout' down in one take. The session was easy-peasy compared to this – and heck, you can't tell Bruno Koschmider you're sick. 'Anyone of us could have been beaten up or killed in Hamburg,' says Tony Sheridan, 'but fortunately, the gangsters and the pimps loved the musicians.'

Because the Beatles returned home unexpectedly, they had no work to hand. Of course, they could play the Casbah and did so on 17 December and New Year's Eve. Allan Williams found them a Christmas Eve dance at the Grosvenor Ballroom in Wallasey. They could have played at his new Top Ten Club in Soho Street, but one night in early December it mysteriously burnt down. Allan Williams had to go to court as the suspicious insurance company refused to pay out. (Another aside: I was in a pub with a Merseybeat group a year or so ago and one of them said, just as you might point out an accountant or a barrister, 'That's Tommy the Torch. He's done more damage to Liverpool than Adolf Hitler.')

The torching of the Top Ten was unfortunate for its new compere, DJ Bob Wooler. He had resigned his daytime job with British Railways and so passed his time in the Jacaranda. Bob Wooler: 'You can write your own entry for *Who's Who* and Paul McCartney has written, "Made first important appearance as the Beatles at Litherland Town Hall near Liverpool in December 1960." It was Tuesday 27 December 1960, a BeeKay (Brian Kelly) dance. I am pleased that I got them the booking, I asked for £8 and Brian nearly collapsed because he was a tight wad – but most of the promoters were. He said he would give them £4 and we compromised on £6, which is £1 a man, five Beatles, and £1 for the driver. I didn't take my ten per cent.'

Brian Kelly added a sticker to the promotional posters, 'Direct From Hamburg – The Beatles'. This didn't say that the Beatles were German, although this is how it was interpreted. Bob Wooler: 'The impact was so tremendous on that Tuesday evening that Brian Kelly posted a bouncer on the door that led backstage to stop any promoter who might be there getting to the Beatles. Brian Kelly signed them to a string of dates for £7.10s, 30 bob a man.' The Beatles played 36 dances around north Liverpool for Brian Kelly in the first three months of 1961.

If Beatle fans could go back in time, that performance at Litherland Town Hall is the one that most would choose. By all accounts, John, Paul, George, Pete and Chas Newby were sensational. Chas Newby?

Oh, I forget to tell you – Stu Sutcliffe stayed with photographer Astrid Kirchherr in Hamburg, so the Blackjacks' Chas Newby played bass. So, if we could go back in time, you now know who the mysterious stranger is.

VIII

Paul McCartney became the Beatles' permanent bass player. He is quoted in the guitar book *Seventeen Watts?*: 'None of us wanted to be the bass player – we wanted to be up front. In our minds, it was the fat guy in the group who nearly always played bass, and he stood at the back. None of us wanted that; we wanted to be up front, singing, looking good, to pull the birds.'

IX

Johnny Guitar was with Rory Storm and the Hurricanes. He kept a diary, which deserves to be published in full with annotations. This is an extract:

'28 September 1960. Got band-jackets from C&A.

29. Got train from Lime Street, missed the other three, caught up with them on the boat train.

30. Arrived Holland 6am, caught the 7am Hamburg Express. Arrived about 5pm. Derry and the Seniors were there, also Beatles.

1 October. Kaiserkeller. We played six hours and finished at 6am. It was hard work. We slept like logs.

2. Had to get up and start playing again. We finished at 5am.

3. We only played four 30-minute sets. The talent contest was a farce, we couldn't understand what people were singing.

4 Indra closed down. Beatles move in with us. Rory, Ringo and I staying at hotel, good one.

9. We started at three, finished at three.

10. Refused to sign extension contract. We get more money, send 80 marks home.

12. Allan Williams came. The Beatles sign their contract so I doubt that they'll get any more cash.

14. Wally and Beatles going to make a test recording tomorrow.'

That heralds a significant event in the Beatles' early history. Pete Best was not well on the 15th and so John, Paul and George met up with Lou Walters and Ringo Starr of the Hurricanes to make a demo at a small recording studio. Wally sang the idyllic lullaby, 'Summertime', and it was the first time that John, Paul, George and Ringo were captured together on record. It's never been heard in public and, as far as I can tell, the only remaining copy is owned by Wally's former wife.

Johnny Guitar's diary gives an overview of the three months from October to December 1960: 'At first hard-going. Rory got notice because he wouldn't stay on stage. The Beatles and us wrecked the stage, Bruno sacked Rory and said we had to pay 65 marks damages. Rory took our big poster. Bruno got the police on to him. Rory got a job in the Top Ten with Tony Sheridan, he worked for a bed, then Beatles got a job in the Top Ten for two weeks and got deported because they burnt down the Bambi Kino. Bruno gave me champagne for my 21st birthday. Gerry and the Pacemakers came from Liverpool and we had a Christmas dinner in the Mambo. We got a job in a fab place when we finished Kaiser but we had to quit because we signed not to play within six months. Two girls saw me off. Rory got home free, we bluffed our way home. New band in Kaiserkeller no good. We picked up piles of souvenirs, mine were smashed. I was sick in the boat, all over the floor, very rough.'

X

Promoter Sam Leach described Pete Best as 'The Atom Beat Drummer'. Pete Best: 'I had to hit the bass drum really hard in the clubs in Hamburg to wake up the crowd. We played non-stop for eight hours sometimes. When we came back to Liverpool, all the groups copied our big drum sound. It was like the birth of Merseybeat.'

Harry Prytherch, drummer with the Remo Four, was on the same bill as the Beatles on 5 February 1961: 'Blair Hall had a stage with a big slope. The first time we saw the Beatles there, the curtains opened and this mighty, mighty sound came out. Pete Best was really hammering the drums and the bass drum started sliding forwards. He was hanging on to his bass drum with one hand and playing it with a stick in the other. He was in real trouble so I got some string and wrapped it around his bass drum pedal and his seat, and every time after that he took a piece of string with him.'

Howie Casey (Seniors): 'There was a Liverpool fashion for drummers to play four in a bar on the bass drum, which is a bit military, Germanic really. Pete used to talk to our drummer Jeff Wallington quite a lot, and George used to talk to Brian Griffiths, who was a wonderfully melodic guitar player. There was a learning curve going on. You're always influenced by other people.'

This is from Pete Best's second autobiography, *The Best Years Of The Beatles*, written with Bill Harry, the editor of *Mersey Beat*:
'When I rehearsed with the band or practised by myself, if I felt something sounds good I decided to develop it. A lot of it was self-taught, plus I always had in the back of my mind Gene Krupa and his big, powerful sound which carried everything.

'In Germany, I was still doing the same thing, but because of the long hours and the fact that we had to develop the music and make it wilder, I began to emphasise what I was doing more. I started slapping the bass drum more to make it a lot stronger, doing more rolls, more cymbal work, and a lot more tom-tom work, which was again a throwback to Krupa.

'Instead of just playing a single or a double-snare drum shot, I started doubling up, so you had this powerful effect. What my right hand was doing on the cymbal, my left hand was doing on the snare drum, basically to emphasise the beat. So you had stages where it was one bang, one bump bump, and you had that fierce bass drum going on all the time, which was like the backbone to it.

'Drummers and other members of bands commented on my style as "this beat which is booming out and surging everything forward." I think it was then, when people began to remark on it, that I began to think about it myself. "What is it that I am doing which is different to them?" Drummers were coming and asking things like, "How do you keep your bass drum going all the time?" Questions like that stumped me. I thought, "Doesn't *everyone* do that?" Then it became apparent, because other people were picking up on it, that maybe there was something special there.'

XI

Peter Cook of the Top Spots: 'Early in 1960, we played with the Beatles virtually every week at the Grosvenor Ballroom in Wallasey. Paul was on rhythm guitar then, he had a Lucky 7, which was a crappy thing. They were a very average band and I used to think that we were far better. Paul did have a brilliant voice for singing "Good Golly Miss Molly" and John had a good voice as well, but apart from that, they were nothing special. They went to Germany and then there was a buzz going round, "Have you heard the Beatles?", and I pooh-poohed it. We played Lathom Hall in Seaforth and the Beatles were on, and I had my arms folded, thinking, "Let's see how good you are." The curtains opened and they started off with "Lucille" and they were so tight and so good that every hair on my neck stood up. I had never heard anything like it in my life. I know now that they had the drums miked up, they had a mike in the bass drum, and they were playing together. Nobody had ever played together before as everybody did their own thing. I was completely in awe of them.'

Alan Stratton of the Black Cats: 'The Beatles were really the first band to have harmonies and they would switch lead vocalists – John would sing one, then Paul and then George which was very interesting as a lot of the bands only had one singer. They were into harmonies, plus that deep, throbbing bass drum with Pete Best. The most exciting thing for me with the Beatles was watching Pete Best set up his drums. The atmosphere was electric. They were intelligent and they knew what was wanted. They would speak to the audience a lot, and if they snapped a string, as Paul did once, he would smile and continue, he wouldn't go off and change it.'

Harry Prytherch, the drummer with the Remo Four: 'Most of us did two or four beats to the bar which was a bomp-ba-bomp, bomp-ba-bomp. Now Pete Best had the original Beatles sound because he would drum eight – bomp-bomp-bomp-bomp-bomp-bomp-bomp-bomp. That was lashing out at you and that was half of their sound. When they went to a recording studio, I think George Martin must have said, "This isn't going to come over on record."'

Wayne Bickerton of the Pete Best Four: 'Pete was a good drummer. All the stories of him not being able to play properly are grossly exaggerated.'

John Cochrane, drummer with Wump and his Werbles: 'I was knocked out the first time I saw the Beatles, not because I thought they were particularly good but because they were playing stuff I liked. We were all trying to be Cliff Richard and the Shadows which is what we thought people wanted, and here was a band that had the nerve to play hard rock'n'roll. There was no *Top Of The Pops* or promotional films then so we had to work out for ourselves how the Americans did it. I used to watch Pete and think, "This doesn't look right but it certainly sounds good." I found it intriguing that he had worked out how to do it.'

Pat Clusky of Rikki and the Red Streaks: 'The Beatles were the first group I heard with the bass drum actually driving along, which made a great rock sound. We'd been pussyfooting around and trying to sound good and the Beatles just came straight at you. That was due to Pete Best, whom I thought was a fabulous drummer and a fabulous person.'

XII

1961 was a good year for the Beatles...maybe they were at their best as a live band. They worked non-stop: in Liverpool January to March; in Hamburg, April to June; in Liverpool, the rest of the year. Hundreds of performances and their first chart hit to boot. The Beatles? A hit in 1961? Does this guy know what he's writing about?

When the Beatles returned to Hamburg in April, they worked with Tony Sheridan at the Top Ten club. Sheridan influenced many Merseybeat musicians, notably John Lennon and Gerry Marsden. I have a video of ITV's *Oh Boy!* from 1959 and Tony Sheridan's stance is pure John Lennon. Les Maguire of Gerry and the Pacemakers: 'Any group that came back from Hamburg would say, "Tony Sheridan, fabulous. He's a knockout." He influenced so many people that the Liverpool sound should be called the Tony Sheridan sound. He did more for Liverpool and the Beatles than anybody else.'

Johnny Hutch of the Big Three agrees: 'It was like going out with an old banger and coming back with a Rolls-Royce. The Beatles owe everything to Sheridan because they copied him to a T. They copied his style on guitar. Sheridan was a fantastic guitarist, the guv'nor.'

Ian Edwards of Ian and the Zodiacs: 'We used to copy a record as best we could, and then we came across Tony Sheridan, who didn't give two hoots as to how somebody else had recorded it. We realised that we shouldn't be carbon copies and we got a lot more adventurous.'

Fred Marsden of Gerry and the Pacemakers: 'I played with Tony Sheridan a few times and he said I wasn't good enough for him. He said I was too slow. He'd say, "Come on, Fred, you're dragging there, get going." He was brilliant but he was unpredictable.'

Tony Sheridan: 'You have a lot of highs, good nights where you turn yourself on and you turn on everybody else. I have decided that being spontaneous is the best way – you've got to surprise yourself. There's no point in playing "Blue Suede Shoes" the same way two thousand times – you've got to be innovative: you put in sevenths, ninths and elevenths, anything to make it interesting. When you're playing several hours a night, you start putting in chords where they don't really belong.'

Despite the accolades, Tony Sheridan never became a star and I didn't understand why until I met him at the 1989 Merseybeatle convention. The first evening he'd had too much to drink and couldn't appear on stage. The next morning he appeared on my radio programme and libelled someone (not a Beatle) who might have been listening (he wasn't, but his mother was). When I said, 'Tony, you don't mean that', he said, 'Yes, I bloody well do.' That evening the nightmare continued as he told Roag Best's band there was no need to rehearse as he'd be playing rock'n'roll standards. He opened with Leonard Cohen's 'First We Take Manhattan', a unique choice for a rock'n'roll standard. I don't know what Tony did the next day but I caught up with him at Southport Theatre two days later. He was again backed by Roag's band but, having spotted Merseybeat friends in the audience, he heckled the band, called them incompetent and invited his mates on stage in their place. 'It was a nightmare,' remembers Roag. 'He would start a song in D and then jump to E. We didn't have a clue as to what he was going to do and I was on the verge of tears.' All this in front of a paying audience who must have wondered what the hell was going on. Okay, he's an excellent musician but I realised that he had no discipline whatsoever. Ideal for a wild, uninhibited place like Hamburg, but not suitable for elsewhere.

In 1961, the German orchestra leader Bert Kaempfert had had a US Number 1 with 'Wonderland By Night (Wunderland Bei Nacht)'. He disliked touring and preferred making sophisticated, orchestral albums with studio musicians. He also worked as an A&R man, discovering talent for Polydor. God knows why, he was giving up a wonderland by night – first-class hotels and restaurants – for Hamburg's underworld. He had wide musical tastes and he signed Tony Sheridan to make records accompanied by the Beatles. He thought German record-buyers might purchase a rocked-up version of an old folk tune, 'My Bonnie', which had been an instrumental hit for Duane Eddy as 'Bonnie Came Back' in February 1960. 'What a silly choice,' recalls Tony Sheridan, 'but Bert Kaempfert said that we had to do something that the Germans would understand and they all learnt "My Bonnie" in English lessons. We went to bed at five and got up at eight to make the record. We took some uppers to get us awake. The guitar solo in "My Bonnie" is all right.'

And so was the drumming. Garry Tamlyn: '"My Bonnie" has a snare rhythm that Pete Best uses consistently on his records and it dates back to the late 1950s. A lot of Sun Records like "Great Balls Of Fire" have that rhythm. You can also hear it in "Surf City" and a lot of surf records. Pete Best's drumming is very tight on this recording. There are snare fills at the end of the four eight-bar phrases, semi-quavers on the snare drum, and they are very tightly executed. It's very accurate drumming on a fast tempo recording.'

The Polydor sessions are chronicled in Discography 1 and you will see that Kaempfert allowed the Beatles to record two numbers on their own. Kaempfert was the only musician to work with the Beatles, Frank Sinatra (writing 'Strangers In The Night') and Elvis Presley (writing 'Wooden Heart', which was based on a German folk tune). 'Wonderland By Night', 'Wooden Heart' and the Beatles – 1961 was a momentous year for Bert Kaempfert.

Bert Kaempfert's instincts were right about 'My Bonnie' – according to *Hit Bilanz*, the book of German chart singles, 'My Bonnie' was on the German Top 40 for 12 weeks, albeit only reaching Number 32. The Germans were the first to put the Beatles on the chart, the Germans had taste, but, mind you, the book also reveals that Sweet had eight Number 1s.

XIII

The Cavern decided to hold lunchtime sessions and because the jazz musicians had day jobs, they had *ipso facto* to be beat ones. Jazz was losing its popularity anyway, and soon the programme was almost exclusively beat music. And, as Bob Wooler says, the rest is hysteria.

The Beatles were working for Ray McFall at the Cavern and other promoters like Sam Leach and Brian Kelly. They filled as many days as possible, usually making the arrangements through Pete and Mona Best. 'She was a real bundle of energy, full of bright ideas,' says Sam Leach.

Southport promoter, Ron Appleby: 'In the early days, that is BE (Before Epstein), the only person you could contact was Pete Best as he was the only one with a telephone. The way we were in those days, anyone with a telephone was rich, they were posh and I think it was also Pete's van. Really, the only way to book the Beatles was via Mona Best and Pete.'

Mrs Best had a tough side. Paddy Delaney, the doorman at the Cavern: 'I wouldn't want to cross Mona Best. She dominated Pete, she was his guiding light, but she had a very nice personality and she liked a laugh and joke whenever she came to the Cavern. She guided the Beatles to a certain extent and took them under her wing.'

XIV

I have spoken to scores of musicians and Cavern dwellers about the Beatles in the Pete Best days. They have their favourite Beatles, but all testify to Pete's extraordinary popularity.

Geoff Nugent (Undertakers): 'It was Pete Best that put the Beatles on the map. Wherever Pete went, you would see two or three girls round the others, and 50 around Pete. You rarely saw him smile, yet he was always pleasant and never nasty. If you look at any of the photos of the Beatles, it is his face that you are drawn to – even if you're a feller.'

Paddy Delaney: 'I remember a show at the Tower Ballroom, New Brighton with a couple of thousand kids and all you could hear was "Pete, smile". Pete very rarely smiled, but there was a trace of humour around the corner of his mouth. If you cracked a joke, he'd give a whimsical smirk and this fascinated the girls.'

Lee Curtis: 'They used to scream like hell for Pete to sing. They wouldn't stop until he sang, and then he did "Matchbox". He used to sing with his head down. It may have been shyness but they loved the way he did it.'

Bill Harry: 'The Beatles even brought Pete to the forefront of the stage with his drums, with the others to the side of him. The girls mobbed the stage so he had to go behind again.'

Pete Best: 'It was Bob Wooler's idea. He said, "It would look good to have everybody together across the stage." When we tried it, the drum kit was nearly pulled off stage and I wasn't too far behind. The other Beatles were in hysterics at the plight I was in.' Really?

XV

Not that all the bookings were the right ones and not that everyone appreciated the Beatles. Take Ken Dodd, who had made the Top 10 in 1960 with the gentle ballad 'Love Is Like A Violin'. He recalls, 'I agreed to do a charity show at the Albany Theatre, Maghull one Sunday afternoon in October 1961. I got there about three o'clock and there was chaos. People were walking out in droves because some idiots on the stage were making the most terrible row. I said, "You've got to get these fellers off, they're killing the show completely" and they said, "Okay, we'll have the interval and then we'll put you on." So they had the interval and while I was changing, one of these idiots came in and said, "Somebody told me that if we gave you our card, you might be able to get us a few bookings." I threw the card away. A year or so later, my agent said, "How would you like the Beatles on your radio show? You can have them for two dates." I said, "I'm not sure. What do you think?" He said, "I think we should just have them for one show because they're going to be one of those groups that fade overnight." We brought them in for one show and Paul McCartney said, "We've worked with you before, Doddy." I said, "No, you've never worked with me, lad." He said, "Yes, we did, at the Albany, Maghull." I said, "That noise wasn't you, was it?" He said, "Yeah, we were rubbish, weren't we?" I said, "You certainly were. I had you thrown off."'

XVI

The show at the Albany had been put together by Jim Gretty, who sold guitars at Frank Hessy's. Hessy's was close to NEMS and if he'd invited the suave Brian Epstein to that concert, would Eppy have appreciated their talent? No, I think he would have taken the Ken Dodd line, so thank heavens he saw the Beatles at the right time in the right context. Luncheon beat at the Cavern.

Bob Wooler had been playing 'My Bonnie' to the troglodytes at the Cavern. Many of them wanted the single but it had only been released in Germany. NEMS was the place to order records.

NEMS boasted that it could fulfil any request so customers would put down a sixpenny (two and a half pence) deposit for a single and Eppy would then track it down. I can testify to NEMS' willingness to find obscure records. I was at the counter one day, a young lad of 16 buying a single by Ben E King or the Drifters, when a man next to me asked surreptitiously, 'Have you got that LP I ordered of Adolf Hitler's speeches?' The record, clearly imported, was produced from under the counter. It amazed me at the time. It amazes me even more now. Brian Epstein was Jewish but he would nevertheless satisfy a customer's request for something as repugnant as this. Business was business.

Eppy certainly knew his customers' requirements. Just a couple of doors away from NEMS in Whitechapel was Beaver Radio, run by Walter Beaver. He is known for three statements: in the 1950s, he said that stereo would never catch on – people wouldn't want second speakers and besides mono was just as good: then, in the early 1980s, he said that video would never catch on as there were already too many repeats on TV and, finally, in the mid-1980s, he came down against CDs, saying that the public would never want to buy their records over again.

As every Beatle book tells you, Brian Epstein's interest in the Beatles began when an 18 year old lad, Raymond Jones, ordered a copy of 'My Bonnie'. I wondered why Raymond Jones had not come forward to claim his part in the Beatles story. Now I know the reason. There was no Raymond Jones.

Alistair Taylor worked in NEMS and was Brian Epstein's personal assistant. He tells me, 'I am Raymond Jones, and Brian Epstein never knew that. Brian would only order a record if there was a firm order in the book. We kept getting asked for "My Bonnie" but no-one had ordered it. I put the release in the order book and made up a name,

"Raymond Jones", and then we set out trying to find it. We rang Polydor in Germany who told us that we had to buy a box of 25. I bought one myself and I told Brian that Raymond Jones had been in for it. Brian put a note in the window saying, "Beatles record here" and within the day, the rest had gone. We kept shipping in boxes of them and Brian rang Polydor in London and suggested that they released it here. They as good as told him to get lost.'

XVII

In July 1961, Bill Harry started a local newspaper about the music scene in Liverpool. He called both the music and the newspaper *Mersey Beat*. By now there were 300 groups on Merseyside so there was plenty to write about, with the Beatles being the predominant local group. NEMS sold the newspaper and Harry asked Epstein to review pop singles and original cast albums on a regular basis. All writers check out their column in print, to see how it looks on the page, to see if there are any misprints and to boost their egos. Therefore, I am certain that Brian Epstein would have looked through *Mersey Beat* to find out what was going on. The newspaper was so heavily Beatle-oriented that he must have seen their name.

Also, the Beatles used to go to NEMS to hear, and occasionally buy, records. They would disrupt staff by requesting repeated plays in the listening booths so that they could scribble down lyrics. More than once, Brian Epstein had told them to move on.

The evidence suggests that Brian Epstein knew who the Beatles were. Maybe not. He appears to have been surprised when he discovered that the group who had recorded 'My Bonnie' were playing regularly at the Cavern, which was only 200 yards away in Mathew Street. He decided to check them out during the lunch-hour on 9 November 1961 with Alistair Taylor, who recalls, 'We looked out of place in our white shirts and dark business-suits. The Beatles were playing "A Taste Of Honey" and "Twist And Shout" but we were particularly impressed that they included original songs. The one that sticks in my mind is "Hello Little Girl".'

As a result of that, Brian Epstein decided that he wanted to manage the group with Alistair Taylor as his personal assistant. His brother, Clive: 'Brian told me about his visit to the Cavern but I didn't take much notice at the time. After all, we had some very successful stores on Merseyside and we were expanding into the suburbs. It wasn't until the spring of 1962 that I took the whole thing more seriously. We had dinner at the 23 Club in Hope Street and he wanted to form a company to manage artists and to promote concerts. At the time I was single and so the idea of some extra activities outside business hours interested me. Also, it was a useful way of diversifying. It became a matter for the whole family as he wanted to use the NEMS name. After careful consideration, my father decided that it was not for him and we decided that NEMS Enterprises should be a company jointly controlled by Brian and myself.'

Their former manager, Allan Williams, is magnanimous. 'I'll be fair and say that Brian Epstein was the best thing that could have happened to the Beatles.'

XVIII

On 31 August 1961, Bob Wooler, in *Mersey Beat*, described the Beatles as the stuff that screams were made of. 'People were always asking me about this group, and I knew they were remarkable, magic even. The only Beatle that I mentioned by name was Pete Best. The girls at the footlights would be looking beyond the front line to this guy on drums, and there was a charisma about him that I found fascinating. The poster for Jane Russell in *The Outlaw* described her as "mean, moody and magnificent", and I applied that to Pete Best.' It stuck.

Pete Best: 'The first time I read it I started laughing, but the fans would say, "It's Pete, he's mean and moody, he must be because it said so in print." I don't think I was. It was more that I was into playing the drums and laying down a rhythm, I'd put my head down and flail away. If it looked like I was mean or moody on stage, it wasn't through me trying to be like that personally. It was a case of putting so much into the music.'

Promoter Ron Appleby: 'When Eppy took the Beatles over, we had a dance at the Kingsway in Southport and we were charging 2s.6d (12 and a half pence) before eight and three shillings (15p) afterwards. Brian Epstein decided that everyone who came into the dance before eight o'clock would be given a photograph of the Beatles. When I saw him at half-past nine, he was hopping mad because the girls were ripping the photograph up and sticking the part with Pete Best on to their jumpers. Pete Best was certainly the attraction with the girls.'

Pete Best: 'I used to be embarrassed about it, to be quite honest. I used to come home and find girls on the path with a sleeping bag or someone looking at you from behind a tree. I didn't know what to do. I'd say, "Hello" or "I'll send you out a coffee in a minute." I took it in my stride and didn't think too much about it. I thought, "If they're doing this for me, then they are doing it for the group." If I got more screams on stage than somebody else, it wasn't a case of that's one for Pete at the back. It was a case of "That's good for the Beatles" – that's all I was concerned about.'

XIX

John Lennon biographer, Paul Du Noyer: 'The stereotype of the rock'n'roll manager is Colonel Tom Parker, who was a complete crook and that seems to be all too common in rock'n'roll management. Even people who don't think that Epstein was a brilliant businessman will always concede that he was a very honest man. He certainly never tried to cheat the Beatles.'

But he did try and change them. Clive Epstein: 'They weren't entirely happy about being put into shirts and collars and suits – this was not the way they appeared in Hamburg, and when Brian sent them to a tailor in Birkenhead, they were far from happy. This was not their style. Nevertheless, it was right because the mums and dads who became Beatle fans liked to see four boys looking clean and tidy in their suits.'

Tony Sanders of Billy J Kramer's original group, the Coasters: 'They didn't need Epstein, he gave them a lot of bad advice – he had them wearing suits, singing the likes of "I Remember You", which wasn't their style and he destroyed the band as it was.'

Bill Harry adds, 'John gave me some photos from Hamburg which included him going on stage with a toilet seat round his neck. As soon as Brian took them over, he asked for them back. He put everything into a plastic image. He sanitised them. They had their hair done at Horne Brothers, got their suits on the Wirral and had Dezo Hoffman pictures. John preferred the raw, aggressive image of the Rolling Stones, and Brian had taken that away from them. I was in the Grapes one night when Brian Griffiths of the Big Three told Epstein to argue with the two lawyers at the end of his arm. Epstein had been giving this wild, aggressive rock band Mitch Murray jingles.' Would John have been happier as a member of the Rolling Stones?

Paul Du Noyer: 'Brian Epstein did a lot for their presentation. I know they hated being put into suits but that seems to be what it took to take them on to the next stage commercially. It did get them a record deal, it did get them to EMI and George Martin. That's where history really starts to move forwards for the Beatles.'

XX

Where did the mop-top hairstyles come from? Okay, it could have been a mop. Eppy might have picked one up from a cleaning lady at NEMS and said, 'That's what I'm going to do with the Beatles.' But I don't think so.

Brian Epstein's friend and Liverpool manager, Joe Flannery, has a photograph of his own mother from the 1930s – she has what looks like a Beatles haircut. 'John Lennon liked my mother very much,' he confides, 'and they based their hairstyles on this photograph.' Again, unlikely.

Watch any 1940s film by those madcap comics, the Three Stooges, and look at their hairstyles. Pure Beatles. Could it be? No, I don't think so.

What about Marlon Brando in the 1953 film *Julius Caesar*? Hamburg photographer Jurgen Vollmer: 'I wanted to look like Marlon Brando in *Julius Caesar* even though I hadn't seen the movie at that time. I was still in school and I told the barber that I wanted my hair just like that, but he talked me out of it. That was the last time I went to the barber – up to this day, I always cut my hair myself.'

Jurgen had his brushed-forward style when he met the Beatles in Hamburg. 'They thought it was funny. I moved to Paris in late '61 and John and Paul visited me there and they wanted to change their clothes. I took them to the flea market and they dressed in the style that I had, corduroy jackets and turtle-neck sweaters. They wanted my haircut so that is when I cut their hair.'

In another variation, Astrid Kirchherr gave the boys their haircuts. 'All my friends at art school used to run around with what you would call a Beatles haircut. My boyfriend, Klaus Voormann, had this style and Stuart liked it very much. He was the first one who had the nerve to get the cream out of his hair and he asked me to cut his hair for him.'

Whatever the reason, Pete Best never succumbed. Astrid Kirchherr: 'Pete's got very curly hair and it wouldn't have worked. Even if he wanted it, I could not have cut his hair that way.'

XXI

Allan Williams: 'I thought the Beatles were a right load of layabouts. I think Eppy only took them on because he was homosexual and they appealed to him.'

Brian Epstein's biographer Ray Coleman: 'There are several areas concerning Brian's private life. Although the homosexuality was crucial to his life, the crucial factor was that he was available for work 24 hours a day. A married man couldn't do that. It was more than just another job to him. He didn't have other priorities and he had money too.'

Sexual freedom was practised, indeed encouraged, around the Reeperbahn in Hamburg and so the Beatles met many homosexuals and transvestites while in Germany. Hence, they would not have been concerned about Brian Epstein's gayness when they signed their management contract on 24 January 1962, although he was acting for them before that date. Epstein was only too aware that the practice was illegal in the UK – even in private between consenting adults – but he was not totally discreet. His personal Menlove Avenue was a bachelor flat at 37 Falkner Street, Liverpool 8.

Alan Sytner, founder of the Cavern and manager when it was a jazz cellar, had known Brian Epstein since his schooldays. 'He got expelled from Liverpool College and several other schools, which is really hard to do. I believe that he was expelled for his homosexuality. His mother told him that the schools were anti-Semetic, but they weren't as I saw a Plymouth Brethren expelled for the same thing. Brian was brought up to believe that he was something special and he developed a superior attitude. He went to school in Cambridge so he gave the impression that he had been to Cambridge.'

An art school colleague of John Lennon's, Ian Sharp knew Eppy's preferences. 'I was involved in amateur dramatics at the time and I could recognise gay people. Billy Hanna introduced me to Brian Epstein, who was definitely camp by Liverpool standards. They were friends and they went to London together, and they also knew Yankel Feather who was outrageously camp and ran a club, the Basement, opposite the Mardi Gras and next door to the Unity Theatre.'

Ian Sharp was talking to John and Paul in the Kardomah café and, a week later, he was surprised to receive a letter from the solicitors, Silverman, Livermore and Co. With Ian Sharp's permission, it is reprinted here for the first time:

'We have been consulted by Mr Brian Epstein who instructs us that on the 21 February last in the Kardomah Café, Church Street, Liverpool, you uttered a certain highly malicious and defamatory statement concerning him to two members of the Beatles.

'We are instructed that in the course of a conversation you said, "I believe Brian Epstein is managing you. Which one of you does he fancy?" The unwarranted innuendo contained in that remark is perfectly clear and is one to which our client takes the gravest possible exception and the damaging nature of which has caused him considerable anxiety and distress.

'He is not prepared to tolerate the utterance of such remarks by you and we accordingly have to require that we receive by return your written apology together with an undertaking that this or similar remarks will not be made by you in the future. We have to make it perfectly clear to you that should we not receive your apology and undertaking, as requested, then our client has instructed us to take such steps as may be necessary to protect his good name and character.'

This revealing letter shows that the Beatles were made aware of Epstein's sexual preference if they hadn't known it already and that Epstein was so annoyed that he was prepared to call Ian Sharp's bluff by resorting to legal action. Also, why should this conversation have been repeated to Eppy and why was Ian Sharp shopped?

Ian Sharp, who works as the actor/director Richard Tate, comments: 'I didn't call Eppy gay maliciously, I was joking about it in an art-college kind of way. I thought that John would have known about it but he must have confronted Eppy with the information. I could have changed the course of musical history by standing up to Eppy and saying, "Bugger you, it's true, I'm standing by what I said", but of course I didn't. I was 20 years old and living with my parents and I panicked. I wasn't used to receiving letters like that and I withdrew the allegations immediately and never mentioned it again. Ironically, years later, I played Sidney Silverman in a production about Craig and Bentley.'

In 1964 Brian published his autobiography, *A Cellarful Of Noise*, but he couldn't discuss a cellarful of boys. Paul Du Noyer: 'He couldn't be honest about his boyfriends in his autobiography, and that is quite touching. Homosexuality was illegal and it would also have been commercial suicide to admit it. It would have taken a very brave man to have proclaimed what was going on in his life.'

There's no doubt that Brian Epstein was deeply attracted to John Lennon, despite, perhaps because of, Lennon's deep sarcasm towards him. 'Sometimes he has been abominably rude to me,' says Eppy in his autobiography. Some Beatle books – notably Peter Brown and Albert Goldman's – have stated that they had a homosexual relationship, which is easy to allege once the participants are dead.

Considering that the female population of Liverpool was drooling over Pete Best, it is likely that Brian Epstein hoped for a relationship himself. I asked Pete about this. He tells me that Eppy propositioned him once on the way to Blackpool: Pete wasn't interested in his stick of rock and the subject was never raised again.

XXII

Monday 5 February 1962.

Pete Best has a virus and Ringo Starr steps in.

John, Paul, George and Ringo play the Cavern at lunchtime and the Kingsway Casino in Southport at night.

XXIII

The *Liverpool Echo* ran a weekly record review column by 'Disker', a pseudonym for Tony Barrow, an old boy of Merchant Taylors' School in Crosby who had moved to London. 'I worked for Decca as the only full-time sleevenote writer in the business. I wrote sleevenotes for Duke Ellington, Gracie Fields and Anthony Newley. Brian Epstein had written to "Disker" and to his great surprise, he received a letter from London. He wanted me to write about the Beatles for the *Liverpool Echo*. I told him I only reviewed records but we should keep in touch.'

Brian Epstein did keep in touch, playing Tony a tape of the Beatles recorded at the Cavern, which does not appear to have survived. 'The sound quality was abominable although it did convey the atmosphere of the Cavern. I spoke on the internal telephone to our marketing department and told them that I had Brian Epstein with me, and they might want to put pressure on A&R for an audition. They didn't know his name but as soon as I mentioned NEMS, they said, "Oh yes, Brian Epstein's group must have an audition."'

The Beatles had their audition in appalling circumstances on New Year's Day 1962, details of which are in Discography 1. Despite below-par performances, the Beatles were convinced that Decca would sign them. 'They're okay, considering we were recording 15 songs in a day,' maintains Pete Best, 'but the German records are of much better quality. They were merchandise and this was just an audition tape. Mike Smith at Decca told us that he couldn't see any problems, so we were shell-shocked when we were turned down.'

Decca preferred the clean-looking Brian Poole and the Tremeloes and Irish balladeers the Bachelors. Epstein tried other avenues for the Beatles and was turned down by EMI. Then a technician who heard the songs at HMV's record store in Oxford Street suggested a meeting with EMI's publishing company, Ardmore and Beechwood, and from there Brian Epstein met George Martin, who ran the EMI subsidiary Parlophone. An audition with the Beatles was arranged for 6 June 1962.

XXIV

The Beatles' first radio broadcast was recorded before an audience at Manchester's Playhouse on 7 March 1962. Alan Clayson, writing in his biography, *The Quiet One – A Life Of George Harrison*, says, 'Afterwards, the four were shocked when they were mobbed by libidinous females not much younger than themselves. Most were after Pete who, pinned in a doorway, would lose tufts of hair to clawing hands while the other three bought their freedom with mere autographs. Despite himself, taciturn Pete was becoming a star. Watching the frenzy sourly was Paul's father, who was to unjustly berate the drummer for stealing the limelight.'

XXV

Former Quarry Man Rod Davis was now at Cambridge University and in 1961 he had made a record for Decca as part of the Trad Grads. 'I bumped into John in Liverpool in March '62 and I said, "I beat you onto record." I told him that I played mandolin, fiddle, banjo, guitar, concertina and melodeon. He said, "You don't play the drums, do you? We need a drummer to take back to Hamburg." That was my second bad career move. My sister remembers my mother saying, "He's not going to Hamburg with that Lennon. He's taking his degree." That's the way she always spoke of him – she never called him anything but "that Lennon".'

Around the same time, Brian Epstein met Rick Dixon, the manager of the Manchester band, Pete Maclaine and the Dakotas. He asked Rick if their drummer, Tony Mansfield, was available to replace Pete Best. Rick said he wasn't but a year later, Eppy took the Dakotas on board as a backing band for Billy J Kramer.

Revealing stories – were John, Paul, George and Brian thinking of sacking Pete Best prior to any dealings with Parlophone? Whatever, he went with them to Germany.

Pete Best: 'We disappeared off to Germany in April and in the meantime Brian had taken a tape of the Decca auditions and was going round all the recording studios and companies in London trying to fix another date or get a contract for us.'

Bob Wooler: 'I was the shoulder Brian would cry on as I had got to know him. He would invite me to the Peacock in Hackins Hey for lunch, which was his favourite haunt, and he would say, "What am I doing wrong? Why aren't they responding?" All I could say was, "I can't believe it, Brian. They should come here and see what the Beatles are doing to audiences." In those days, A&R men didn't hurry to a provincial town to see a group. It was different once the Beatles happened – we had a rush of A&R men up here.'

Pete Best: 'While we were there in Germany – in fact, we opened the Star-Club – we received a telegram from Brian saying, "Congratulations – you've got a recording test with EMI. Get cracking, write your own material. I've got some material back home which they want you to practice." That was when they wrote "Love Me Do". We wanted to do our own songs anyway. We'd done standards, we'd done the American releases and other bands had copied us, and we realised that doing our own songs would be the next trick. We were dead cocky about it. I did dabble in writing some lyrics myself and I've still got them.'

Tony Sheridan: 'We'd all signed the contract with Polydor, not just me. Bert Kaempfert came to some arrangement with Brian Epstein and the silly bastard let them out of the contract for nothing.'

XXVI

Brian Epstein told many people that the Beatles were going to be bigger than Elvis. Not too much should be read into this – how many managers say their clients are going to be bigger than Elvis and then neither the artists nor the managers are heard of again? It is hype. If he really believed that, why did he sign so many other acts?

Norman Kuhlke, the drummer with the Swinging Blue Jeans, recalls, 'Brian Epstein may have said that the Beatles were going to be bigger than Elvis but when he tried to sign us up, he said, "I like your music the best. The Beatles don't play my kind of music." He was more jazz-minded than rock-minded.'

Clive Epstein: 'I remember Brian coming back from London with the acetate of "Love Me Do" and he played it to the family and me. We liked it but when he suggested that these boys would be really enormous – he saw them as legends and bigger than Elvis – I couldn't quite see it…but how right he was.'

Billy J Kramer comes down on Eppy's side: 'Brian Epstein said that the first time he heard the Beatles he knew that they were going to be as big as Elvis. Well, the first time I saw them at the Litherland Town Hall, I thought that, and friends of mine thought I was crazy. But I really knew it, I was positive that they would do it.'

XXVII

Parlophone was a joke label, literally. Its big-selling albums included 'At The Drop Of A Hat', 'Songs For Swingin' Sellers' and 'Beyond The

Fringe' and its artists included Charlie Drake, Spike Milligan and Bernard Cribbins.'

We now know that Parlophone was the Cinderella label of the EMI group, but in June 1962 it was the ugly sister. The popularity of its biggest act, Adam Faith, had nose-dived. George Martin: 'I was pretty desperate for anything, and I envied Columbia, which had so many hits with Cliff Richard. I saw Brian Epstein and heard his tapes. They weren't very impressive but there was something peculiar about the way they sounded that I thought should be looked into. I asked Brian to bring them down from Liverpool so that I could have a look at them. I was immediately impressed by them as people, not particularly as musicians.'

EMI recording engineer Harry Moss: 'The Beatles came here in 1962 for an artist test in Studio 3 and that was the first time I met them. Around that era, before anybody was put on to tape to be recorded by a big organisation, they did a test first and I attended hundreds of these tests. I was blasé about it because of every hundred that you had to sit through and suffer, there would only be one that was any good. Frankly, I wasn't impressed by the Beatles at that time.'

At the audition, they played their stage favourite, 'Besame Mucho' and two songs written by John and Paul, 'Love Me Do' and 'PS I Love You'. In 1995, the original version of 'Love Me Do' was included on 'Anthology 1' and at last we could hear what George Martin heard. It is slower, rougher and has less sparkle than the hit single. Pete Best sounds as though he's banging bin-lids and his drumming goes off on a strange tangent in the middle. If this was all George Martin heard of Pete Best, is it all that surprising that he was sacked?

Pete Best: 'George liked us and said, "Fine, rehearse 'Love Me Do' a little more and come back and we will put the finishing touches to it." At some time between that session and the next, I got kicked out.'

George Martin: 'I thought Pete was an essential part of the Beatles because of his image: there was a moody James Dean look about him. But I didn't like his drumming. I didn't think it held the Beatles together as it should have done and I was determined that the Beatles weren't going to suffer because of it. I told Brian that I was going to use a session drummer when we made records. I didn't realise that the other boys had been thinking of getting rid of Pete anyway and that my decision was like the last straw that broke the camel's back. So Pete was given the boot, poor chap. It was hard luck on him, but it was inevitable.'

I DID SOMETHING WRONG

'What was the name of the drummer sacked by the Beatles
and replaced by Ringo Starr?'
University Challenge, April 1996 – no-one knew the answer

I

Brian Epstein's confidant, Peter Brown, writes in his biography, *The Love You Make*: 'George Martin was particularly critical of Pete Best's heavy, uninventive drumming. When the audition was over, the most George Martin would say was "Maybe". In John, Paul and George's minds, Pete Best was already doomed as he sat next to them in the van on the way back to Liverpool.' Dramatic, but not very accurate.

Because of their Hamburg engagement, the Beatles had not played Liverpool for two months. From 9 to 20 June 1962, they performed exclusively at the Cavern and recorded a second session for the BBC. For the radio programme, *Here We Go*, they performed 'Ask Me Why', which would end up as the B-side of 'Please Please Me', their old favourite 'Besame Mucho' and a contemporary hit, Joe Brown's 'A Picture Of You'.

On 21 June they performed at the Tower Ballroom on a Bob Wooler production, starring Bruce Channel and Delbert McClinton, the hit recorders of 'Hey! Baby'. 'Hey! Baby' with its catchy harmonica riff was an influence on 'Love Me Do', although Frank Ifield puts in a claim for 'I Remember You'. The Beatles performed 'Hey! Baby' and 'I Remember You' from time to time, both with John playing harmonica.

Bruce Channel recalls, 'It was a tremendous audience at that ballroom. I'd had a Number 1 record but I was surprised that so many people were there. Then I found out that the Beatles were on the bill and they went on just before me. John Lennon liked Delbert's harmonica playing very much. I can remember him talking to Delbert quite a bit and Delbert was showing him how he played. I liked his harmonica on "Love Me Do" very much and I also liked the song.'

Leo Sayer: 'I was fascinated by the harmonica on the Beatles' "Love Me Do" as I'd never heard anyone play a harmonica like that before. All I knew was Sonny Terry and Brownie McGhee who were all (demonstrates) chugga-chugga-chugga, just a rhythmic thing. John Lennon was using the harmonica to make a riff and what he was playing was more like a trumpet line. I played the harmonica myself and this was a very good influence on me.'

Several Beatle books state that the EMI recording contract was kept secret from Pete Best and yet the poster for the Tower Ballroom event states that the Beatles are 'Parlophone Recording Artistes'. Didn't Pete Best see the advertising?

II

The weeks went by – the Beatles' appearances centred around the Cavern, but there were some odd appearances such as the Barnston Women's Institute (30 June 1962) and the *Royal Iris* on the Mersey River (6 July and 10 August). The Grafton Rooms (now known for its 'Grab a Granny' nights) succumbed to rock'n'roll on 3 August, with a bill featuring the Beatles, Gerry and the Pacemakers and the Big Three. Brian Epstein also presented two shows with the Beatles and Joe Brown and the Bruvvers, and Joe recalls, 'I was one of the first persons from outside Liverpool that Brian Epstein actually booked. He booked the hall, put on the show and tried to make money out of it. I remember that we followed the Beatles. Normally, if you're topping the bill, you finish the show: if you're second top, you close the first half. Brian wanted the Beatles on just before me, hoping that the fans would be screaming so much that we wouldn't be able to get on at first. It was a good trick and the Beatles were a real hard act to follow. We did it, mind you, to our credit we did it.'

Local musician Steve Kelly was at the Joe Brown/Beatles show at the Cambridge Hall in Southport. 'I'd gone to see Joe Brown and I didn't go into the hall for the Beatles. Someone told me that they were a greasy, noisy, sweaty band and I wasn't missing out on anything.'

Granada TV had seen the Beatles on that showcase and decided to film them at the Cavern for their *Know The North* programme. The filming would take place on 22 August and it would be the Beatles' first TV appearance. In all probability, it precipitated Pete Best's sacking as the Beatles wanted to be seen with a new drummer.

III

Much to everyone's surprise, Mona Best was pregnant. Musicians wondered who the father was as regular visitors to the Casbah had never seen Johnny Best there. It is common knowledge among Merseyside musicians that the father was the Beatles' road manager, Neil Aspinall, although this has never been revealed in any Beatles book. When the Beatles returned from Germany in June 1962, Neil left his regular employment as a trainee accountant and became their full-time roadie. He was very friendly with Pete Best and had lodgings at Hayman's Green with the Best family.

Roag Best was born on 21 July 1962. 'At the time I thought Roag was Pete's full brother,' says Beryl Marsden, and many other musicians thought the same. Johnny Guitar of Rory Storm and the Hurricanes: 'I got on very well with Mona Best and so did Rory, but I never met Johnny Best. Pete told Rory the truth, and Rory told me that Neil was his father when Mona Best was pregnant.'

The birth was registered by Mona Best on 31 August 1962 and gives her son's name as Vincent Rogue Best (*sic*). She lists herself, Alice Mona Best, as the mother and John Best as the father. In keeping with the times, the birth was made legitimate.

But there was no announcement of the birth in the *Liverpool Echo*.

IV

3 August 1962, according to Philip Norman's *Shout! The True Story Of The Beatles*: 'August began, and still Pete Best knew nothing of the contract with Parlophone. The Beatles, en route for the Grafton in West Derby Road, were all in Mona Best's Oriental sitting-room, waiting for Pete to come downstairs. He did so in high spirits, full of the Ford Capri car he had almost decided to buy. Mrs Best remembers that Paul, in particular, showed unease over the price Pete intended to pay for the car. 'He went all mysterious. He told Pete, "If you take my advice, you won't buy it. You'd be better off saving your money."'

V

In the early 1960s, teenage girls preferred their idols to be unencumbered. Managers would tell the stars to keep their girlfriends from the press and their marriages secret, if they couldn't be persuaded from taking such a ridiculous step. Joe Brown recalls, 'If it got out that a pop star was married, his career could be ruined. The idea was to be untouchable but always available. On a pedestal but on the same level as the fans, if you know what I mean. My manager didn't want anyone to know that I was getting married and he made me wear a black wig. The idea was to whip it off when I got to the altar, but I'd pencilled my eyebrows and I realised that I'd look very silly with a blond crew-cut and a black eyebrows. I kept the wig on and Vicki started arguing with me at the altar. The press hounded me for months. They were sure that I was married but they couldn't prove it. They made my life hell and they tried all sorts of tricks to find out. They'd ring up my mother and say they were the jewellers wanting to know about the alterations to the wedding ring. One night two reporters turned up at my house at half-past ten in the pouring rain, clutching a copy of our marriage certificate. My manager told me to make a statement at ten o'clock the next morning. My publicity agent rang round all the papers but only two of them turned up. The press were getting a bit sick of this nonsense by then. Fortunately, the Beatles killed everything like that.'

But not at first. Brian Epstein told John and Cynthia that the Beatles' popularity would be harmed if they were seen as a couple and, for its time, it was sensible advice. Cynthia says, 'I was in a horribly vulnerable position. The fanatical Lennon followers did not take kindly to me. I was a threat to their fantasies and dreams. The most dangerous place for me was the ladies' toilet. Sometimes I thought that I wouldn't get out in one piece. My solution was to keep a very low profile and keep my mouth firmly shut. I was no match for all those girls.'

Early in August 1962, Cynthia told John that she was pregnant. 'It was such a shock to both of us that I was pregnant. It was anything but a celebration at the time as John had to tell Mimi and I had to tell my mum.' Even worse, they had to tell Eppy.

In 1962, if you got a girl into trouble, you married and John immediately said that they should get married. The wedding was arranged for 23 August at the Registry Office in Mount Pleasant, the day after their TV appearance, and John would spend his wedding night playing at the Riverpark Ballroom in Chester. Start as you mean to go on.

John was getting married, going to be a father and he had to write songs for the Beatles' recording session. Writing about Pete Best in his autobiography, Brian Epstein said 'He was friendly with John, he was not with George and Paul.' If there was to be a spirited defence to retain Pete, John would be too preoccupied to give it.

VI

On Wednesday 15 August 1962, the Beatles played lunchtime and evening sessions at the Cavern. This marked Pete Best's final appearances with the group – and, indeed, the last time he would ever speak to any of them.

Pete Best says, 'I'd been with the Beatles for two years. We'd been through thick and thin together. There were times when the money from bookings wasn't enough to keep things going. There was a strong fellowship about the group and I never thought that they wanted to get rid of me. On Wednesday night when we'd finished, Brian said he'd like to see me in his office the next morning. This was quite normal because, with the family phone, I fixed the bookings and he'd used to ask me about venues and prices.'

Bob Wooler: 'I learnt that Pete Best was going to be sacked on that night, not before. I could imagine it with someone who was constantly late or giving problems, but Pete Best was not awkward and he did not step out of line. I was most indignant and I said, "Why are you doing this?" but I didn't get an answer.'

Pete went to bed a happy man. Eppy had a sleepless night.

VII

It was not a foregone conclusion that any drummer, if invited, would want to join the Beatles. Norman Kuhlke of the Swinging Blue Jeans: 'A lot of drummers wouldn't have wanted to join the Beatles. I was having such a good time in the Blue Jeans, I don't think I would have wanted to change groups. The Beatles were just another Liverpool group at the time.'

In the *Anthology 1* video, Paul McCartney says, 'We started to think that we needed the greatest drummer in Liverpool.' Was that automatically Ringo Starr?

The Beatles were very impressed with Joe Brown's drummer, Bobby Graham, formerly with Mike Berry and the Outlaws. 'Brian Epstein invited us back to the Blue Angel after a show. He called me to one side and told me that he was having problems with Pete Best's mum and he wanted him out of the Beatles. He asked me if I would take his place. Although I liked the Beatles, I turned him down because I didn't want to come to Liverpool. Besides, I liked Joe Brown, who was having hit records. I met George Harrison about five years ago and he had no idea that Brian had asked me.'

Strangely enough, Bobby's path kept crossing with the Beatles. 'George Martin used me for session work and I did get involved in one of the Beatles' early sessions although I've no idea which one. When Ringo had his tonsils out, I was asked to take his place on tour for a few days, but I was getting so much session work that I couldn't do it and I recommended Jimmy Nicol instead. I did play with the Beatles on one of their *Pop Go The Beatles* sessions from the Paris Theatre and that was because the BBC didn't think Ringo was adaptable enough for what they wanted at that time.'

From Gerry Marsden's autobiography, *I'll Never Walk Alone*: 'Whatever they planned for the Beatles, Pete certainly didn't figure in it, which was a tragedy of a kind. They asked my brother Fred to play with them, go to Hamburg with them, but he told them that he had decided to stay with me, the biggest mistake he'd made in his life!' A nice quote, but Fred Marsden tells me not to believe everything I read in books. 'Also, I could never have had a Beatles hairstyle. They'd have looked stupid with William Hague on drums.'

When Mike McCartney was interviewed by Libby Purves about a children's book on Radio 4 in December 1992, he added that he had once been a drummer with the Quarry Men and that he would have replaced Pete Best in the Beatles, had he not broken his arm. Why did he feel after 30 years that he had declare this? Did it really happen? To be fair, there was a set of drums in the McCartney household and both Paul and Mike had practised on them. Also, Mike had played a snare drum on an early Quarry Men home recording of 'One After 909'.

Fred Marsden continues: 'The only person in Liverpool who did drum solos in Liverpool was Johnny Hutch from the Big Three. He was a very good drummer and even in 1962 he was into a heavy sort of rock music rather than pop. Technically, he was the best in Liverpool – well, he was the only one who did solos so he must have been the best.'

Johnny Hutchinson: 'Bob Wooler said, "Well, Brian, I think John would suit the Beatles down to the ground." Brian said, "I do too. What do you think, John?" I said, "I wouldn't join the Beatles for a gold clock. There's only one group as far as I'm concerned and that's the Big Three. The Beatles couldn't make a better sound than that and, anyway, Pete was a very good friend of mine and I couldn't do the dirty on him like that, but why don't you get Ringo? Ringo's a bum – Ringo'll join anybody for a few bob." Ringo was playing at Butlin's with Rory Storm, who had signed a contract to play there. Ringo just got up and left – "Bugger you, boy, I'm going to higher things." He had no scruples at all.'

If there was any possibility of Johnny Hutch joining the Beatles, they were dispelled by the few gigs he'd play in the period between Pete's sacking and Ringo's arrival (see page 46). There was considerable friction between Hutch and Lennon, and quite clearly Hutch was not prepared to be subordinate to him.

Bob Wooler: 'The Beatles didn't want a drummer who would be a force to be reckoned with and hence, Johnny Hutch didn't stand a chance. Trevor Morais (of Faron's Flamingos) was also considered but he was a centre of attraction and they didn't want all the showmanship. They wanted a very good drummer who would not intrude and Ringo played that role very well indeed.'

VIII

How it happens in fiction:

Mike is the Stray Cats' manager in *Stardust* (Fontana Books, 1974) by the Liverpool author, Ray Connolly: 'Mike looked at him and hated him more than ever. The grovelling little bastard, he thought. But he smiled. "Johnny! Fancy a drink?" he said, and with an arm round Johnny's shoulders, he led him away to a sudden and merciless slaughter. He just couldn't afford to let some little two-faced twat like Johnny interfere with his plans now. No way.'

That isn't far from real life. Mike Middles tells how the lead singer of the fledgling Durutti Column was sacked in *From Joy Division To New Order – The Factory Story* (Virgin, 1996): 'That night, Tony Wilson visited the flat of singer Phil Rainford. To make matters worse, Rainford seemed unusually enthusiastic, painfully exclaiming in an increasingly excited tone about his plans for the band and how marvellous the recent rehearsals had been. As every excitable comment passed by, Wilson found himself sinking deeper and deeper into despair. He sat, quietly panicking, preparing to administer the chop while a Bruce Springsteen album filled the room. "At the end of side one," thought Wilson, "I'll tell him then." Inevitably, side one cluttered to a halt and side two began to spin threateningly. Wilson decided to tell him at the end of side two, and so he did. Feeling profoundly wretched, with Rainford's tone of disbelief ringing in his ears, he strode wretchedly away from the startled singer's flat. Tony Wilson had tasted the darker side of band management.'

And so to the Beatles...

When he awoke on the morning of Thursday 16 August, Pete Best put on his T-shirt and jeans and asked the Beatles' roadie, Neil Aspinall, if he wanted to come into town with him.

They drove to Whitechapel and Pete went into NEMS while Neil waited outside. Pete went into Brian Epstein's office, sat down and, as he says, 'It took just ten minutes to change my life forever.'

Less than that actually because Epstein simply said, 'The lads don't want you in the group anymore.' Not "the lads and I", not "we", not "I", but "the lads". Brian Epstein was distancing himself from the decision.

That may be right. Brian Epstein recognised Pete's popularity and liked him a lot. He would offer him another drumming job a few days later – the fact that he didn't do at this meeting implies that this was a hurried decision, that he hadn't sorted everything out. He also expected Pete Best to work with the Beatles for the rest of the week.

Pete Best told Bill Harry in *The Best Years Of The Beatles* (1996): 'There was a phone call while I was there and when he answered it, Eppy said, "I'm still with him at the moment." I don't know who phoned, it could have been anyone. I wasn't paying too much attention to who was phoning, as I was still trying to fathom the situation.'

The sensible money is on McCartney. Well, it must be – some years earlier Pete Best had spoken to Philip Norman for *Shout! The True Story Of The Beatles* (1981): 'While I was standing there, the phone rang on Brian's desk. It was Paul, asking if I'd been told yet. Brian said, "I can't talk now. Peter's here with me in the office."' So, why the reservation this time round? Surely, though, if Pete was sure it was Paul, he would have grabbed the phone and told him where to go.

A young Liverpool band, the Merseybeats, were waiting outside to see Brian Epstein. He was about to sign them. They saw Pete Best emerge looking as though he had seen a ghost and Epstein in tears. Eppy told them to make another appointment.

Back on the street, Pete met up with Neil and went for a drink. Pete says that by chance, they bumped into Lou Walters from Rory Storm and the Hurricanes, but I wonder if it was so coincidental. Bobby Thomson had temporarily replaced Wally in the Hurricanes for the Butlin's season and we'll come to Wally's possible role in a minute.

IX

Ringo Starr – Richard Starkey – the oldest of the Beatles had been born in the Dingle in 1940. He had had a traumatic childhood with one illness after another and, not surprisingly, he left school with no qualifications. In 1957 he joined the Eddie Clayton Skiffle Group and then, in 1959 and owning a full drum-kit, he became part of Rory Storm and the Hurricanes. He worked for an engineering works, Henry Hunt and Sons, and he was encouraged to pack it in for a season at Butlin's holiday camp in Pwllheli. Rory was an excellent showman but only a moderate vocalist. In order to add some glamour to the band, Rory insisted that everyone in the band should play a leading role and he introduced a solo spot, 'Ringo Starr-time'. Ringo would sing undemanding pop and R&B songs of the day including the Shirelles' 'Boys' and Johnny Burnette's 'You're Sixteen'. He was a pleasant, rather than a good, vocalist but he was highly-rated as a drummer.

Harry Prytherch, drummer with the Remo Four: 'Ringo was a lot more technical than Pete Best. There were five or six of us who liked discussing the technicalities – Ringo, Kingsize Taylor's drummer, Sonny Webb's drummer, myself and Billy Buck out of the Jaywalkers – and Pete Best was different from us, there's no doubt about that. You noticed the difference when Ringo took over because Pete was a real pounding rock'n'roll drummer.'

Fred Marsden, drummer with Gerry and the Pacemakers: 'I knew Ringo years before he joined the Beatles. He was always listening to records and getting to grips with the technical side of drumming. That's why the Beatles wanted him. Ringo only lived a quarter of a mile from me in the Dingle and after an afternoon session at the Cavern he would watch us or the Beatles even if he wasn't playing himself. We would then go back and listen to records.'

Dave Lovelady, drummer with the Fourmost: 'I think Ringo would admit that he has never been a brilliant drummer technically, but he had a very unique drive and he was very good to watch. He used to throw his head all over the place and the beat that he produced was really pounding. He had a very unusual style but technically he wasn't brilliant.'

Bobby Thomson, guitarist for Kingsize Taylor and the Dominoes and later the Rockin' Berries: 'Ringo used to set the kit for a right-handed drummer even though he was left-handed. He could play so evenly with either hand, he was such a rock-steady drummer that once he started a tempo he never moved. A similar drummer is Bev Bevan from ELO, a real bricklayer and I mean that in the nicest sense. On the other hand, Trevor Morais of Faron's Flamingos and the Peddlers was a very flowery, flashy drummer and I don't think I could play with him in the same band, there would be too much going on. Roy Dyke, who became part of Ashton, Gardner and Dyke, was also very good but he was very jazz-influenced which is fine if you're playing that type of music.'

Johnny Guitar of Rory Storm and the Hurricanes: 'We had a great band in Skegness in 1962. There was Bobby Thomson, Ty O'Brien, Ringo and me. To be honest, we couldn't wait for Rory to take his break so that we could get into some hard instrumental rock'n'roll.'

Ritchie Galvin, drummer with Earl Preston and the TT's: 'Ringo was wasted with Rory Storm because although Rory was a great showman, he was a dire singer. No wonder Ringo said yes.'

43

X

Billy Butlin opened his first holiday camp in Skegness in 1936. The concept was to provide mass, on-site entertainment and catering for the British working man and his family. The rows of chalets looked like army barracks and the Redcoats organised the camp with military precision. The holidaymakers were even told when to get up with a voice over the tannoy saying 'Wakey-wakey, campers'. The holidaymakers enjoyed themselves in the swimming pool, playing billiards, darts or tennis, old-time dancing, watching variety shows, at the funfair or simply making new friends. The Skegness camp incorporated a zoo which attracted national publicity in 1962 when an elephant fell into the swimming-pool, upturned and drowned.

By the early 1960s, Billy Butlin realised that the camps were losing their appeal. The British public was becoming more free-spirited – they wanted to holiday abroad, good heavens – and older teenagers no longer wanted to holiday with their parents. He introduced rock'n'roll nights to attract adolescents and the advertisements implied that sexual freedom was the order of the day. 'Wakey-wakey' was the call to move back to your own chalet.

In 1962 Rory Storm and the Hurricanes were booked for the summer season at Skegness. The Beatles would never have accepted such a booking. John Lennon referred to the camps as 'Belsen'.

XI

Rory Storm and the Hurricanes and the Beatles were good friends. Holiday camps apart, they often worked the same venues and had spent several weeks together in Hamburg. They got on well with each other and Ringo had sat in with the Beatles on occasion.

Mind you, Ringo had sat in with many bands – he had worked with the Seniors and had left the Hurricanes in January 1962 to be part of Tony Sheridan's backing group at the Top Ten Club. Although he was back with the Hurricanes, he was unsure about his career and was considering emigrating to America.

Kingsize Taylor and the Dominoes were in Hamburg with their drummer, Dave Lovelady. Dave recalls, 'After we'd been in Hamburg for two months, the time came when I had to come home and return to my studies. Teddy Taylor and the rest of the boys wanted to stay professional, so it was decided that I would leave and they would fly out a replacement. Teddy wrote to Ringo to ask him if he'd like to take my place. He wrote back to say that he would and he gave Rory Storm his notice.'

So, in August 1962, Ringo Starr had no intention of joining the Beatles or even of returning to Liverpool and his girlfriend, Maureen Cox, in the near future. He was going to finish at Skegness and then join Kingsize Taylor and the Dominoes. Their promised £20 a week was good money, but the Beatles would offer £25.

XII

There are conflicting stories as to how and when Ringo Starr was invited to join the Beatles.

Version 1 – According to Mark Lewisohn's *The Complete Beatles Chronicle*, John Lennon telephoned Ringo at Skegness on Tuesday 14 August, two days before the dismissal. This sounds reasonable but is fraught with difficulty. The chalets did not have telephones and getting through to anyone at a holiday camp in 1962 was a time-consuming and usually fruitless task. Also, Ringo was with Johnny Guitar in a caravan just outside the campsite.

Version 2 – In the *Anthology 1* video, Ringo Starr recalls, 'It was a Wednesday and Brian Epstein called; I don't remember John coming over, which is in somebody's book. "Would you join the band, really join the band?" I said, "Sure, when?", and he said, "Now". I said, "I can't do that, I'll join you on Saturday." We had Saturdays off as that was when they changed the campers. So I gave Rory Thursday, Friday and Saturday to bring someone in.' This is more plausible than Version 1 because Ringo was invited to join the Beatles as an employee (£25 a week rather than a split of the takings) and also because a well-spoken businessman would have more success in getting the holiday camp to locate Ringo.

Version 3, the one Ringo doesn't remember, has John Lennon visiting Belsen. Johnny Guitar remembers, 'John and Paul knocked on the door of our caravan about ten o'clock one morning, and I was very surprised because John hated Butlin's. Paul said, "We've come to ask Ringo to join us." We went into the camp and Rory said, "What are we going to do because this is mid-season and we can't work without a drummer?" Paul said, "Mr Epstein would like Pete Best to play with you." We couldn't stand in Ringo's way 'cause we knew the Beatles were going to be big. We went back to Liverpool and saw Pete, but he was so upset that he didn't want to play with anybody.'

I asked Johnny Guitar if he was sure he was correct, that John and Paul did visit Ringo at Skegness. 'Yes, Rory got a big shock when Ringo said he was going to leave, and so did I. It is possible that Ringo had been tipped the wink on his last visit to Liverpool, but we had no inkling of what was going on.'

Travelling the 170 miles from Liverpool to Skegness in 1962 was no joke. A train journey would involve changes and take several hours. John Lennon couldn't drive and Paul would probably want a relief driver. Quite possibly, that would be Neil Aspinall. There were no motorways and a likely route would be leaving Liverpool on the A580 to Manchester. From there, it is on to the A628 to Marple, A57 to Sheffield and keeping on that road to Worksop and Lincoln. Then it is leaving by the A158 to Wragby, Horncastle, Spilsby and, finally, Skegness. Even maintaining a speed of 30 miles an hour, it is unlikely that the journey could be done in less than five hours. No matter how early they set off on the morning of Thursday 16 August, there could be no guarantee that the Beatles would return in time for a show at the Riverpark Ballroom that evening. And with Ringo in tow.

But what if it happened on Tuesday 14 August as the Beatles weren't working that day and they could take Neil Aspinall as well? With a full datebook, isn't it more likely that the Beatles would line up their replacement before they sacked Pete Best? Mightn't Ringo Starr think it was grossly unfair to Pete Best and turn them down? In any event, they weren't to know that Ringo was about to join the Dominoes.

I think that John, Paul and possibly Neil went to Skegness on the Tuesday 14 August and saw Ringo, in spite of his comments. That paved the way for the sacking of Pete Best. Possibly Rory contacted Lou Walters, the Hurricane who was still in Liverpool, and asked him to sound out Pete about joining the Hurricanes. However, Wally found Pete so depressed that he realised that it was neither the time nor the place to invite him to join the Hurricanes. Hence, the Hurricanes returned on Saturday and Pete turned them down.

XIII

This would also explain why Brian Epstein wanted Pete Best to play a few more dates with the Beatles. He knew that Ringo couldn't join until Saturday. Once Pete had gone, he rearranged the meeting with the Merseybeats and set about finding a temporary replacement drummer. Fortunately for him, Johnny Hutchinson of the Big Three agreed. (He had already played with the Beatles at a Larry Parnes audition in May 1960; see page 12.) This time, he was to stand in on Thursday 16 August 1962 at the Riverpark Ballroom and on Friday 17 at the Majestic Ballroom in Birkenhead and the Tower Ballroom, New Brighton. Pete Best was said to be 'indisposed'.

Johnny Hutchinson: 'I was playing with the Beatles and the Big Three at the same time. I would play the first half-hour with my group and get dressed up in my band suit, set the drums up and do half-an-hour with them, unset my drums, take the suit off, shoot off elsewhere and do half-hour in the Beatles gear and go back for half-an-hour with the Big Three.'

Hutch also offers this gem. 'Brian asked me to bear with him for a few more weeks. The Beatles were even going to get a fantastic drummer from Leeds. This chap came down from Leeds – he was about 54, balding and very big, not at all as Brian expected and Brian had to start hiding. Apparently, he turned out to be no good as well. That's when I started playing with them.'

XIV

When he sacked Pete, Brian Epstein was worried that Neil Aspinall might resign in sympathy. As it turns out, his initial reaction was to resign but Pete told him to continue, although it would mean leaving Hayman's Green. No-one has spoken of the tensions in Hayman's Green, but it can't have been easy for Pete. He was telling his half-brother's father to leave the house. When Neil turned up to take the Beatles to Riverpark that evening, he asked the Beatles why Pete had been sacked. 'It's got nothing to do with you,' said John. 'You're only the driver.'

Not even Neil – or Nell as the Beatles called him – may know that Brian Epstein had considered a replacement for him – John Booker, who later worked with the Undertakers. 'I was going round with the Merseybeats at the time and Eppy came up to me with George Harrison and said, "John, I'd like you to look after the Beatles." I said, "You're joking." He said, "No, I'd like you to take over from Neil." I said, "No, I wouldn't do that to Neil." That showed me what Eppy was like.'

XV

On Saturday 18 August 1962, Ringo Starr joined the Beatles as a full-time member for a horticultural society's annual dance at Hulme Hall, Port Sunlight. The real test would come the following evening among the regulars at the Cavern club.

The Cavern's doorman, Paddy Delaney: 'George Harrison had gone downstairs – there weren't many people in at the time and there was a bit of a commotion. I went down to see what the trouble was and George was holding his eye. He had a beauty of a black eye.'

George Harrison in the *Anthology 1* video: 'The Cavern had three tunnels and I stepped out of the dressing-room into a tunnel and some guy butted me in the eye.'

Ringo Starr admits in the same video, 'We played the Cavern and there was a lot of fighting and shouting – half of them hated me and half of them loved me.'

Mike Gregory of the Escorts: 'Everyone was screaming at Ringo and throwing tomatoes at him. They were shouting for Pete Best and giving him a hard time.'

Ian Edwards (Ian and the Zodiacs): 'I was very friendly with Ringo and I felt very sorry for him at the time. They were shouting "Ringo never, Pete Best forever" and refusing to let them play. There was a big question as to whether this could be the Beatles' downfall. Everyone was talking about it. Ringo had been playing in a group which wasn't taken seriously and suddenly he's in the biggest thing on Merseyside. He was a very good rock drummer, but there were a lot of better drummers around, such as Johnny Hutch.'

Ray Ennis (Swinging Blue Jeans): 'It was murder when Pete got the sack. George Harrison got a black eye and there was a big split in the Beatle fans, they were fighting each other. Pete got a raw deal and without doubt the luckiest man alive is Ringo Starr – and yet I've never heard him say that.'

Ritchie Galvin of Earl Preston and the TTs: 'I really wanted the Beatles to do well as I thought it might open the floodgates a bit. We might get away from groups like Shane Fenton and the Fentones. I was a bit surprised when they sacked Pete and I hoped they hadn't blown it.'

Diana Mothershaw, who sold records at Rushworth and Dreaper's: 'Shortly after Pete was sacked, John and George came into Rushworth's when it was quiet. We asked them why Pete had been sacked and they said he couldn't drum well enough. He only had his own style. Then someone told them that Pete was only round the corner looking at drums and they said, "See you, girls", and ran out. We shouted "Cowards" after them.'

From Brian Epstein's autobiography, *A Cellarful Of Noise*: 'The sacking of Pete Best left me in an appalling position in Liverpool. Overnight I became the most disliked man on the seething beat-scene. True, I had the support of the Beatles who were the city's darlings and they were delighted to have Ringo. But the fans wanted Pete Best as a Beatle and there were several unpleasant scenes.'

XVI

The truth – Pete Best: 'I felt like putting a stone around my neck and jumping off the Pier Head. I knew that the Beatles were going places and to be kicked out on the verge of it happening upset me a great deal. I was sure we were going to be a chart group. For weeks afterwards, I just wanted to forget about everything, I didn't want to see the drums, I didn't want to see people. The fact that they weren't at my dismissal hurt me a lot more than the fact that Brian told me that I wasn't a Beatle any longer.'

The spin-doctored version appeared in *Mersey Beat*, which was published on 23 August 1962: 'Pete Best left the group by mutual agreement. There were no arguments or difficulties, and this has been an entirely amicable decision.'

This version is so far from the truth that it casts doubts upon Bill Harry's integrity as an editor. Was he merely Epstein's mouthpiece? Was he only the person on Merseyside who thought it was an amicable split? He says, 'I was working an 80-hour week on *Mersey Beat* and I took Brian Epstein's word for some things I should have checked out. When he gave me the story about Pete leaving, he made it seem like a mutual agreement. I can see now that I was being manipulated.'

But surely Bill had known of the chaos at the Cavern and surely fans had been calling his office? 'Yes, we were inundated with calls but we had already gone to press. I never doubted Brian Epstein's veracity when he gave me that story. We had to go to press three or four days beforehand – we were only a little job for the printers, Swale's, who also published the *Widnes Weekly News*.'

Mersey Beat had gone to press before the full facts were known and being a fortnightly paper, the dismissal was old news by the time of the next issue as Ringo Starr was securely established as the Beatles' drummer. Bill Harry: 'I felt an injustice was being done, but not because Pete was getting kicked out on the brink of success. That's the luck of the game. I felt that there should have been some truth about why he'd been put out. They should have said, "We've decided that we get on better with Ringo, and we want Ringo with us." Instead, they suggested that Pete Best wasn't good enough.'

Philip Norman's *Shout! The True Story Of The Beatles* says that Pete's fate was decided in a pub meeting between Paul, George, Brian Epstein and, not John Lennon but Bob Wooler. There may be some substance in this. 60 year old Ted Knibbs was Billy J Kramer's first manager. Epstein bought his contract for £50 of which Ted got £25. He remembered a meeting shortly after the sacking, 'Brian Epstein, Bob Wooler and myself were having a few drinks in the New Cabaret Club and Bob said, "I'm going to tell the whole Pete Best story in *Mersey Beat*. I think the fans should know why Pete Best was dropped for Ringo." Brian went into a right flap – "You won't, you won't" and Bob said, "I will, I will" and this went on and on. I said, "I hope there'll be no trouble" and Brian stood up, called the waiter, paid the bill and said, "I'm sorry, gentlemen, but I must leave your company." If he'd had a cloak, he would gathered it about him and stalked out. I said, "Now you've done it, haven't you, Bob?" He said, "Why shouldn't I tell the story?" I said, "Because you're not a journalist, mate, you're not relying on that for your living. You're relying on Brian Epstein who employs you to put on his shows." He said, "Well, what do you advise me to do?" I said, "First thing in the morning, go round and tell Brian that you are sorry about last night and get it all over with."'

Bob Wooler says, 'I was annoyed about what happened to Pete Best because I couldn't see any reason why he should have to leave the group. People said he wasn't a very good drummer – well, it makes you wonder who is a good drummer these days because Ringo wasn't even on the first record. But I was an outsider looking in. I was going to write an article called 'Odd Man – Out' but it never materialised and I regret that very much.'

Did Bob Wooler repair his relationship with Brian Epstein? 'Yes, sense prevailed and I made it up with the Nemperor. Ted Knibbs, who was older and wiser than me, said I had made my Declaration of Independence and that was enough.' Strangely enough Bob's concept of the odd man out was taken up by Albert Goldman, a writer not normally known for his perception. He wrote, 'The odd man out had to get out – and the new man in had to be an odd fellow.'

But would the story have ever been published in *Mersey Beat*? Bill Harry: 'Once Brian Epstein came into the picture, I felt we were being manipulated. We had a reporter in Widnes who was a Billy J Kramer fan. He would pick up the proofs and show them to us, usually on a Saturday. He also took the proofs to show Epstein, but we didn't know he was doing that until we caught Epstein with them one day.'

XVII

More spin-doctoring took place on 23 August 1962 when John Lennon and Cynthia Powell were married. Paul, George and Brian Epstein attended the ceremony, but not Aunt Mimi. At the wedding breakfast at Reece's Restaurant opposite Clayton Square, Brian Epstein told them they could live in his flat at 37 Falkner Street. Brian had used the flat for his furtive trysts and one of the first songs John wrote in the flat was 'Do You Want To Know A Secret?'. There were a lot of secrets in the Beatles – Brian's love life, John's wedding and why they sacked Pete Best.

Cynthia Lennon: 'We were only children in those days, 20 and 21, and we were used to being told what to do and what not to do, so it wasn't too much of an effort to keep our marriage a secret. That's the way it seemed to be at the time, according to the pop world. Pop stars weren't supposed to be married so we fell in with this thing. It was no problem to me as I knew that I had my man and I loved him and he loved me and we were having a baby. Whatever anybody else thought didn't really matter to me.'

XVIII

Pete Best: 'In the weeks that followed, Brian got in touch with me and said, "There's another band I'm interested in promoting. Would you like to play drums for them? I'd like you to join the group and build it up into another Beatles." I said, "I'm flattered that you don't want to lose me, but because of the vicious way it happened, the backhanded way it took place, I can't agree to come back and let you be my manager." He said, "Okay, but the offer's still there whenever you feel like it." In the meantime, Joe Flannery came down to my house and asked me to join Lee Curtis. He knew he wasn't going to make a lot of money out of it but he wanted me to join the band and be a part of the team. I thought about it for a couple of days and then rang him up and said, "Okay". I didn't want to cash in on being an ex-Beatle and I often spoke to Joe Flannery about this when I saw, "Pete Best, the ex-Beatles drummer" on posters.'

As it happens, Brian Epstein had even set that up. He had asked Joe to contact Pete with a view to forming Lee Curtis and the All Stars. Lee Curtis: 'I was so delighted that he came and joined me because he was such an asset to us. Pete Best drumming behind you was a tremendous attraction, it really was. Joe came up with the name the All Stars as they were all stars in their own right from different bands.'

Unfortunately for Pete Best, Lee Curtis was the wrong vocalist for 1963 – his highly overcharged performances were too late for the Elvis Presley era and too early for Tom Jones.

Beryl Marsden also worked with Lee Curtis and the All Stars. 'We used to rehearse at Peter's house and his mum would be telling tales of how awful the Beatles were and how I musn't go near them, never speak to them, or listen to their music. I thought it was a terrible thing for them to do, but I liked their music and I couldn't stay away too long. I rang up one day and said, "I'm not very well. I can't come in and rehearse," and I skived off to a lunchtime session at the Cavern. Unfortunately I got caught out and got a really bad scolding from Peter's mum, but it was worth it.'

Pete Best: 'We played on the same bill as the Beatles on two occasions. One was at the Cavern when we were second on the bill to the Beatles. The other was in the *Mersey Beat* Pollwinners Concert. On both occasions we were on just prior to the Beatles, and we had to pass one another...face to face, yet nothing was ever said.'

XIX

By joining the Beatles, Ringo Starr had set up a chain reaction.

Rory Storm and the Hurricanes were Ringo-less at Skegness. Johnny Guitar: 'Someone at the camp said he'd play drums for us. We said, "Are you a drummer?" and he said, "No, I'm an actor." He was Anthony Ashdown, who had been in *The Loneliness Of The Long Distance Runner.* He got us by for a week or two until a relief drummer could come out.'

Drummer Dave Lovelady was returning from Hamburg and Ringo was going to replace him in Kingsize Taylor and the Dominoes: 'About ten days after the first letter, Kingsize got a second letter from Ringo saying that he was joining the Beatles instead. I came home and we did a swop with the Four Jays. Brian Redman took my place, and I joined the Four Jays or the Fourmost, as they were to become.'

Brian Redman stayed five months with the Dominoes and then joined the country-rock band, Sonny Webb and the Cascades, who became the Hillsiders. Kingsize Taylor: 'We then took Gibson Kemp to Hamburg, who was playing for Rory Storm at that time. As he was only 15, I had to go to London and sign a guarantee to get him out of the country: I had to act as a guardian for him. He was a cracking drummer, the greatest rock drummer who ever came out of Liverpool. I was so glad I took Gibson and everybody seemed to be happy about the changes right the way down the line.'

Except perhaps Rory Storm. Lee Curtis: 'Rory Storm told me that he'd got another new drummer. I said, "Why another one? Gibson Kemp's a knockout." He said, "I've lost Ringo and now I've lost Gibson as well as he's joining the Dominoes." I said, "What's going on, Rory? Why can't you keep your drummers?" He said, "What can I say? I make 'em and they take 'em."'

ASK ME WHY

'What do you call someone who hangs around with musicians?'
'A drummer.'
Musicians' joke

WHAT'S THE REASON I'M NOT PLEASING YOU?

In this chapter, I discuss the possible reasons for Pete Best's sacking and hopefully reach a satisfactory explanation. This is not as easy as it sounds because even Pete Best doesn't know why. 'I wish I knew the answer. It would put a lot of heart-searching to rest. I may have been getting too much attention but it didn't matter a dicky-bird to me. They said I was anti-social and non-comformist but each individual member had his own following and, when you added it together, the following was fantastic.'

Gerry Marsden praises the Beatles throughout his autobiography, *I'll Never Walk Alone*, except when it comes to the sacking of Pete Best. 'I thought it was a tacky thing to do for no apparent reason. I was very annoyed that when I'd asked Brian Epstein for a reason, he couldn't give me a proper answer. I thought it was a sour way to start a recording career for the Beatles, firing a drummer who'd been with them for two years. I told Pete Best – who wasn't particularly a mate of mine, but was an honest feller who got a bad deal – what I thought, but by then it was too late. The deed had been done. Ringo proved perfect, but the principle of Best's sacking left a nasty taste in the mouth as the Beatles began their climb out of Liverpool to the world stage.'

THE EX-FILES 1
Pete Best Was A Lousy Drummer

First, the pro-Pete Best lobby. Excuse the repetition, but I want to emphasise that many musicians who heard Pete play with the Beatles had no reservations about his ability.

Johnny Guitar of Rory Storm and the Hurricanes: 'I'd seen Pete playing for over two years with the Beatles and I'd never known anyone complain about his drumming.'

Drummed Out!

Brian 'Noddy' Redman, drummer with the Hillsiders: 'There was no other group like the Beatles at the time and they were destined for the top all the way along. Pete Best was fine for the Beatles, but then nobody had seen any other drummer with them. At the time, we just accepted him so we were surprised when he was booted out.'

Wayne Bickerton, later to be part of the Pete Best Four: 'Pete was a good drummer. All the stories of him not being able to play the drums properly are grossly exaggerated. There was nothing wrong with Pete's drumming.'

John McNally of the Searchers: 'I was surprised by the story that the record company didn't think Pete was good enough. I somehow doubt it because he was superb. He was doing what all the punk bands did later – he was the first to do fours on the bass drum, bom, bom, bom, bom, which gives more power, as opposed to ba-bom, ba-bom, the jazz-type drumming. It was powerhouse drumming with loads of cymbals and he was great. Ringo and Chris Curtis also did similar things and if you listen to the Star-Club albums, you can hear that bass drum thump through everyone's act.'

Chris Curtis (drummer with the Searchers): 'You could sit Pete Best on a drumkit and ask him to play for 19 hours and he'd put his head down and do it. He'd drum with real style and stamina all night long and that really was the Beatles' sound – forget the guitars and forget the faces, you couldn't avoid that insistent whack, whack, whack.'

Earl Preston: 'The Beatles had a unique sound and Pete contributed a lot to it. It was very, very raw rock'n'roll, a very fast driving beat that other drummers tried to emulate but never could. He had a unique style, he used both hands at the same speed. Most drummers play four with the cymbals and then one with the snare, but he doubled up so that both hands were doing the same rhythm which was a very effective, terrific sound.'

Beryl Marsden: 'I didn't see why the Beatles had replaced their drummer. When Pete joined our band, his drumming was great and he was also a great bloke.'

Steve Fleming of Mark Peters and the Silhouettes: 'Although we knew the Beatles well, they weren't very intimate or forthcoming about internal matters. All of a sudden when Brian Epstein took over, Pete Best disappeared. It was a tragedy as the Beatles sounded better with Pete Best than they did with Ringo Starr.'

We'll dampen this praise with a note of caution from Brian Epstein's personal assistant, Alistair Taylor. 'Brian told me that the other Beatles had come to him and told him that they wanted a different drummer. It was hard for Pete who had gone through all those years of rather dodgy venues with them, but I always told Brian that Pete's drumming was a touch uneven and that he didn't quite fit into the group.'

THE EX-FILES 2
Pete Best Was Not A Versatile Drummer

So far, we can conclude that Pete Best was not a lousy drummer, but there are indications that he was not a versatile one. He was a one-trick pony and Jackie Lomax of the Undertakers puts it succinctly: 'Pete Best could only play one drum beat, either slowed down or speeded up.'

Brendan McCormack, classical guitarist, formerly with Rikki and the Red Streaks: 'The role of the drummer wasn't clear at that time. The vocalists were the important people, the bass didn't really count and it is only in the 1980s that the bass came into focus. Drummers simply had to keep the band playing in time. It's changed now, it's changed radically. You should not view the sacking from today's perspective.'

Fred Marsden, drummer with Gerry and the Pacemakers: 'Pete Best was excellent for the Beatles – I don't think you could have found a better drummer for the material they were doing, which was mostly Coasters-type stuff. But as they advanced, I don't think he was technically good enough. I felt sorry for Pete Best because he was an excellent drummer in his field.'

Billy Hatton of the Fourmost: 'His bass drum technique was four to the bar – bom, bom, bom, bom – which initially gave them that thumping sound, but when they started doing stuff that required more sophisticated drumming techniques they needed a better drummer.'

There was no doubting the ability of the Beatles' bass player. Bob Wooler: 'Paul McCartney was the outstanding bass player on Merseyside. Perhaps Johnny Gustafson had the edge on him, but Paul was exceptional.'

Paul's ability contributed to Pete's downfall. Garry Tamlyn: 'There was a very close association between the drummer and the bass guitarist in rock'n'roll bands. A bass guitarist would tend to base what he was doing on what the drummer was doing and forge a very close ensemble with the drummer, so Paul would pick up any fluctuation in tempo very quickly.'

This could cause friction. Bob Wooler: 'The Beatles used to play the Cavern at lunchtime and sometimes they would stay behind and rehearse, and just myself and the cleaners would hear them. One day Paul showed Pete Best how he wanted the drums to be played for a certain tune and I thought, "That's pushing it a bit."'

Maybe, but it was typical behaviour. Ritchie Galvin: 'Sometimes after a lunchtime session in the Cavern, we would spend the afternoon in the Mandolin Club in Toxteth. Paul was showing Pete the drum pattern that he wanted on a particular song. Pete tried to do it but he didn't get it. He did argue quite a bit with Pete, and Paul was a frustrated drummer, which is unusual as so many drummers are frustrated front-liners. He always made for the drums on jam sessions at the Blue Angel – Gerry Marsden would be singing and Wally Shepherd would be playing guitar.'

From this, we conclude that McCartney was the Beatle who was most aware of Best's limitations and, unfortunately for Pete, he was also the Beatle who wanted to expand their repertoire. While the Beatles remained a rock'n'roll band – and John Lennon, to some extent, was always a rock'n'roller – there were no problems with Pete Best's drumming. A contrast can be made with the drummers who worked in New Orleans for Little Richard and Fats Domino. They came from a jazz heritage and weren't allowed to produce ornate drumbeats.

If Pete Best was not a good enough drummer, could he have taken lessons and improved? Garry Tamlyn: 'Rock'n'roll drumbeats are quite simple in comparison with jazz beats for example, and so, given time, it doesn't automatically follow that a rock'n'roll drummer would be able to do that. From what I've heard, I can't gauge how talented Pete Best was so I don't know if he would have been able to cut things as well as a practised session drummer.'

It is argued that Pete Best would not have been suitable for what the Beatles did later; the strange time signatures of 'Strawberry Fields Forever' come to mind, but that was new music and who's to say what's right and what's wrong? Ringo, as it were, wrote the book and became a very stylistic player. However, in 1962, Lennon and McCartney had no idea that they would be writing songs as innovative as that. Their musical aspiration was to be the next Goffin-King, ie to produce carefully-crafted pop songs along the lines of the Brill Building writers in New York. As Kingsize Taylor says, 'There was no big change in their repertoire when Ringo joined, at least not at first, and so it was still the same songs.'

At McCartney's instigation, the Beatles had added some middle-of-the-road ballads to their act, which can be viewed as a throwback to dance-band days. Ritchie Galvin, drummer with Earl Preston and the TTs: 'Pete was a very basic drummer and not very technical. The Beatles, particularly Paul, were singing songs like "Till There Was You", "A Taste Of Honey" and "Fever", songs that called for a little more from the right hand than ticky, ticky, ticky. The bass drumming needed for those songs is quite intricate.'

The big show song from the Liverpool groups has to be 'You'll Never Walk Alone' from Gerry and the Pacemakers, and how sophisticated is the drumming on that? Garry Tamlyn: 'Even a show song like "You'll Never Walk Alone" when played by Gerry and the Pacemakers is only triplet rhythms on cymbals, and that's no different from what Earl Palmer was doing in the 1950s with Fats Domino and Lloyd Price. Any competent drummer could play that. You're only calling the song sophisticated because of the emotions it contains. It doesn't require any complex drumbeats.'

THE EX-FILES 3
George Martin Didn't Like
Pete Best's Drumming

Paul McCartney told Mark Lewisohn: 'When we first came down in June 1962, with Pete Best, George Martin took us aside and said, "I'm not happy about the drummer." And we all went, "Oh God, well, I'm not telling him. You tell him...Oh God!" and it was quite a blow. He said, "Can you change your drummer?" and we said, "We're quite happy with him, he works great in the clubs." And George said, "Yes, but for recording he's got to be just a bit more accurate."'

George Martin was unhappy with Pete Best's drumming but he hadn't suggested that Pete should be sacked. He could make the Beatles' records with a session drummer and no-one need know – who knew, for example, that Gary Walker didn't drum for the Walker Brothers but read comics in the control booth? Gary was kept in the Walker Brothers because he looked good, a reversal of the Pete Best situation. Then again, Dennis Wilson, was forced into the family group, the Beach Boys, at the insistence of his mother. He was only a moderate drummer and was often replaced by session men, but he became essential as the heart-throb of the group.

However, the very fact that it was George Martin who said Pete Best wasn't good enough may have been a deciding factor. The Beatles were desperate for a UK recording contract and George Martin was a London-based, authority figure with an impressive CV as a producer. His opinions would carry more weight with the ambitious Beatles than Brian Epstein's. George Martin's views were, at the very least, the catalyst for the sacking of Pete Best. Hamburg record producer Paul Murphy: 'I believe Pete Best was a casualty of George Martin. Getting that record contract was like getting the Holy Grail. It was like "We are not worthy" when they went in the doors at EMI, so I do think George Martin had a lot to do with it.'

Ironically, there could be another dimension to this. Was George Martin good enough for Pete Best? Playwright and Cavern dweller Willy Russell: 'I don't think the recording industry could cope with the weight of the beat that the Beatles were capable of in those days. When we did *John Paul George Ringo...And Bert*, we included the Beatles' recording of "Long Tall Sally", but we overdubbed another bass line to try and capture the powerful bass sound that they had at the Cavern. George Martin never got near the way they could do rhythm'n'blues, and even "Twist And Shout" sounds thinner than it did live. I get irked when people say the Rolling Stones was the great rock'n'roll band: for me, the Beatles' beat section – the bass and drums – was much stronger live than the Stones.'

Harry Prytherch of the Remo Four, is of a similar mind. He feels that George Martin may have found a dominant bass drum seeping through the mikes of the other performers. Garry Tamlyn: 'Charles Connor, who was in Little Richard's touring band, wasn't allowed to record with him for that reason, but that was in the 1950s. The studios were more advanced by the early 1960s. The drums in the Pete Best tracks for Parlophone are fairly mixed back, although the timbre of his executions suggest he was hitting the drums pretty hard. He was a hard and forceful drummer.'

THE EX-FILES 4
Ringo Starr Was A Better
Drummer Than Pete Best

Actually, Pete Best was given a reason for his sacking, but he chose not to believe it. 'Brian Epstein said it was because I was not a good enough drummer, but that has never held water with me. Most of the drummers in Liverpool copied the style I had brought back from Germany. I was faster than Ringo, but otherwise we were similar.'

Pete adds, 'I had known Ringo for years. We met and talked when the Casbah first opened, and Ringo used to play with Rory Storm. We became good·mates in Germany and we knocked around in Germany and also when we came back to Liverpool. I wouldn't rate Ringo as a better drummer than me. I'm adamant about that and lots of people would support me.'

But certainly not Ringo Starr. 'Ringo said I wasn't a good enough drummer and a few other things in a *Playboy* interview. It took a long time but we got a settlement out of court and an apology as well.'

Nevertheless, Ringo's still saying it. In a tetchy *Q* interview in June 1992, he said, 'Did I ever feel sorry for Pete Best? No. Why should I? I was a better player than him. That's how I got the job. It wasn't on personality. It was that I was a better drummer and I got the phone call. I never felt sorry for him.'

Paul McCartney backs him up. 'I wasn't jealous of Pete because he was handsome,' McCartney told Hunter Davies. 'That's all junk. He just couldn't play. Ringo was so much better. We wanted him out for that reason.'

In *Many Years From Now*, Paul gives an example. 'There was a style of drumming on "What'd I Say", which is a sort of Latin R&B that Ray Charles' drummer, Milt Turner, played on the original record, and we used to love it. One of the big clinching factors about Ringo as the drummer in the band was that he could really play that so well.'

That Ringo Starr was a better drummer than Pete Best is the most obvious reason, the easiest to understand and the one Pete was given, but is it true?

Trevor Morais, drummer with Faron's Flamingos and, later, David Essex: 'There was an immense difference between them. Ringo was one of the best drummers in town whereas Pete Best was just good for the job. I don't know why they changed as Pete was ample for the job and very popular, but Ringo was an excellent drummer, no doubt about that.'

Bobby Thomson of Kingsize Taylor and the Dominoes: 'Pete was a good drummer, but he wasn't as solid as Ringo and he was a little too heavy on the bass drum. All you could hear when Pete played was the bass drum always four to the bar, but he was very good and I saw no reason for him being sacked. I preferred Ringo as a drummer but that is a personal preference. I felt so sorry for Pete – some people would have committed suicide.'

THE EX-FILES 5
The Other Beatles Were
Jealous Of Pete's Good Looks

The concept as to who is or isn't good-looking has changed with the years and it is, in any event, a subjective matter, depending on personal taste. Few people in the early 1960s thought Mick Jagger was handsome, but his looks became acceptable. On the other hand, Brian Epstein regarded Billy J Kramer as one of the most handsome men in the world, but his looks seem too bland, too conventional now. I don't think that anyone would deny Pete Best's good looks, but did he look that much better than Paul McCartney or John Lennon?

Beryl Marsden: 'Pete was definitely the handsome one of the crew. Paul was baby-faced, George was not very attractive, and John had his own kind of look. Pete was a really attractive guy and I'm sure there was a lot of jealousy on their part.'

Kingsize Taylor: 'I think Pete Best was tossed out of the group because he was too good-looking. You would see the Beatles on stage and all the girls wanted to see was Pete Best. Pete was a swarthy, curly-headed feller and he looked much better than McCartney.'

Geoff Taggart: 'Pete Best was a fine drummer. He had style and he was a hell of a good-looking bloke. He reminded me of the actor Jeff Chandler who was very popular at the time. When they played the Plaza in St Helens, I noticed that all the girls stood at the front of the stage and were watching Pete Best. The other three were doing all they could to get their attention but he was the only one that they were bothered about.'

Ray Ennis (Swinging Blue Jeans): 'If George Martin had seen a live performance, he'd have discovered that Pete was the star. When he came to the front to sing – and he couldn't sing very well – they would scream at him. They used to tell the other Beatles to sit down so they could see Pete at the back.'

Mark Peters: 'We used to go bowling with the Beatles and the groups late on a Saturday evening at the Tuebrook Bowling Alley. There was rivalry between us but also a lot of friendship, especially on Saturday nights when we used to play bowls until four in the morning. The story went round that Paul and John thought that Pete Best was far too handsome, that was the joke at the time, but there seems to be a lot of evidence to show that this was true.'

Ian Edwards of Ian and the Zodiacs: 'Pete Best was the star of the group, he was the good-looking lad that all the girls went for – but I never thought it would come to that. I was very surprised when he went.'

Ritchie Galvin of Earl Preston and the TTs attributes the decision to Brian Epstein: 'It was Pete Best's misfortune to be such a good-looking boy: the management wanted the front-line to have the fans and so there was in-fighting going on.'

This is supported by Gerry Marsden in his autobiography, *I'll Never Walk Alone*: 'Musically, perhaps Ringo was slightly better than Pete Best. But the change wasn't necessary for that reason, in my opinion. I was pretty sure it was a political firing, which sprang from Pete being too handsome. He certainly attracted the girls and I think Brian saw his good looks as some kind of threat to Paul and John and George as they were beginning to climb to national fame. Hence, they got Ringo!'

The argument that after Pete Best the Beatles chose someone ugly is not wholly true – a frog to take the place of a prince, as it were. (And it was those frogs that suffered in Macca's childhood.) Like Mick Jagger, Ringo had his own look and Eppy must have been very amused when the press said that Ringo looked Jewish. Ironically, Ringo grew to resemble Yasir Arafat. Also, Ringo with his nodding head was very entertaining to watch. Unlike the sullen Pete Best, who kept his head down, Ringo looked as though he enjoyed playing the drums. George Harrison admitted as much in a 1962 letter to a fan which was auctioned at Sotheby's 20 years later: 'Ringo is a much better drummer and he can smile – which is a bit more than Pete could do. It will seem different for a few weeks, but I think that the majority of our fans will soon be taking Ringo for granted.'

Jimmy Tushingham, who took Ringo's place – and wore his stage-suit – in Rory Storm and the Hurricanes: 'Pete Best was a good drummer and I reckon he got pushed out of the Beatles because he was such a good-looking guy. Everybody liked Pete Best – Ringo was a nice feller but he wasn't that cool.'

Pete's mother became Moaner Best when it came to discussing Pete's dismissal. 'I asked Brian what the reason was and he said, "I can't tell you." I said, "Well, may I give you my reason? It's jealousy, Brian, because Peter is the one with the terrific following." I think that Peter had to be dismissed at that stage because, if they became nationally known, Peter would have been the main Beatle with the others just the props.'

In a 1980 interview with Patrick Doncaster entitled 'I Still Don't Know Why I Was Sacked'(!) in the *Daily Mirror*, Pete Best said, 'It was jealousy possibly, not that I was aware of it. But I was getting more fan reaction and some people had been tagging us Pete Best and the Beatles.'

It may be more than just fans screaming at you. Earl Preston is certain that Pete's looks attracted 'a tremendous number of women to the band', and the Beatles had large sexual appetites. Did Paul resent the number of women who fell into Pete's arms? Certainly, Paul wasn't the number one pin-up from the Beatles, but when you read the large number of casual flings in his quasi-autobiography, *Many Years From Now*, you wonder if he could have found time for any more.

Also, by 1962, both Pete and John were dating their girlfriends seriously and both of them were warned against marriage by Brian Epstein, who thought it might detract from their fan following. John, however, had few thoughts of fidelity. Ritchie Galvin of Earl Preston and the TTs: 'I remember one afternoon at the Mandolin Club in Toxteth. Paul was on the grand piano singing Sinatra songs and everyone else was lounging around in the big armchairs they had there. A couple of girls were there and one of them decided to suck John's dick in front of us. John was talking about the lunchtime gig while all this was happening.'

I have a feeling that either all three Beatles resented Pete's brooding beauty, or none at all. If only Paul or George were jealous of Pete's good looks, they would have been teased mercilessly by John. For this argument to hold, John must have been jealous of Pete himself.

THE EX-FILES 6
Pete Didn't Fit In With The Beatles

Pete Best had the fan base, but he could never claim to be the leader of the Beatles. George was young and inexperienced and the decisions were made by John or Paul, more usually John. John often got his way through belligerence, whereas Paul McCartney was the PR man of the group. They shared decisions about their repertoire, but McCartney had to put up with Lennon's banter if he was forced to play something against his wishes. Bob Wooler says, 'They often stood as Harrison, McCartney and Lennon on stage, and that was Paul's favourite position as it made him seem the leader.' Nevertheless, during the first taped interview with the group on 27 October 1962, Paul McCartney acknowledges John as the leader of the group.

Liverpool singer and Hamburg record producer Paul Murphy: 'John Lennon was the leader of the Beatles. He had the stage presence and body language and he looked in control. Lennon's attitude always was, "We're going to be the biggest band in the world." I remember meeting him at the Oasis Club in London, Shaftesbury Avenue, which was an all-night sauna and coffee bar, and he was talking about how his band was going to make it. He had a very powerful personality and he could appear dictatorial.'

Al Peters: 'There's no doubt that John Lennon was the leader of the Beatles. Paul McCartney had a lot to say but Lennon had it in attitude and presence. Pete seemed a loner compared to the rest of the Beatles, but it can be like that in a band. You can't all be buddies.'

Ian Sharp: 'I thought John Lennon was totally outrageous. I told him that he'd either end up in jail or be someone famous.'

Lennon's impulsive, drunken wit could taunt people. Bob Wooler: 'The setting is the Blue Angel and Paul McCartney is upstairs talking to some press people, while in the basement is John Lennon shooting his mouth off, well away with drink or whatever. He said, "Hitler should have finished the job", meaning that the gas ovens should have been more active than they were. His manager was Jewish and I prevailed upon him to be quiet because the press were upstairs but he didn't take any notice of me. I told Paul that John was shooting his mouth off and that the press must not get wind of it. That was an example of John's indifference. He enjoyed the danger associated with some of his remarks, and of course he did say "We're more popular than Jesus now". It's on the cards that he made the Hitler remark to Brian, which certainly would have offended him, but Brian would have let it ride as he hated flare-ups. It was a terrible thing to say, even as a joke, and I put it down to his lackadaisical upbringing.'

Ritchie Galvin of Earl Preston and the TTs: 'There was a floating beat night one night on the *Royal Iris* with Earl Preston and the TTs and the Beatles. We shared the captain's cabin as our dressing room and John was helping himself to the whisky in the captain's rack. Girls were coming in for autographs and you know what ship doors are like. One girl had her hand on the jamb and John just kicked the door on her hand and laughed. No-one else laughed and the girl's hand was dripping with blood. To be honest, I never liked him much.'

Lennon – and it is surely significant that so many musicians refer to him as 'Lennon' rather than 'John' – was indifferent to others, and many people didn't like his sarcasm and arrogance. Earl Preston: 'John was the first punk, in his attitude towards people. He was an educated Scouser and, if you put the combination together, you have someone who really believed in himself.'

Norman Kuhlke of the Swinging Blue Jeans: 'John Lennon had his own circle of friends and I never spoke much to him. I thought that he had been instrumental in getting Pete out of the group, but I may be wrong.'

Pat Clusky of Rikki and the Red Streaks: 'John could be very quiet. If he didn't want to speak to you, he wouldn't. George was very shy, whereas Paul exceptionally fancied himself, he really thought he was something. I liked Pete the best. He was a smashing bloke, great to talk to.'

Norman Scroggie of the Lee Eddie Five: 'In their own ways, both Paul McCartney and John Lennon were aloof, and Pete Best was always quiet and kept to himself. On the other hand, George Harrison was the most pleasant of the four. If you had a guitar in your hand, he'd talk to you for hours.'

Paddy Delaney, the doorman at the Cavern: 'John could turn on you and he could be very short with anyone. He'd have a go at you if needs be. Paul was quiet, and he was called the Choirboy by the girls. George was the quietest of the lot, saying nothing until he was spoken to but he had a kind heart. Pete Best was also very quiet but he was the most popular of the group because he was so good-looking.'

Bill Harry: 'I always liked Pete Best. He never said anything, he was the most difficult person to interview. He'd sit by himself and he didn't really fit in with them as a personality. When we were in the Jacaranda, they'd sit and talk and he'd be in the corner by the window. If somebody talked to him, he'd just grunt or whatever.'

Vocalist Cy Tucker: 'The birds used to go crazy on Pete Best but he was very quiet and laid-back. He was not as full of life as the other three guys. Pete Best wasn't doing their thing.'

I shouldn't think it was easy to fit into a group with John Lennon, and Pete Best, if the truth be known, did well to last two years. Paul and George did fortunately share some of John's sense of humour. Paul McCartney told Mark Lewisohn, 'Pete Best had never quite been like the rest of us. We were the wacky trio and Pete was perhaps a little more sensible; he was slightly different from us, he wasn't quite as artsy as we were. And we just didn't hang out that much together. He'd go home to his mum's club, the Casbah, and although we'd hang out there with him, we never really went to other places together.'

Above: Ringo in Rory Storm days with band-mate Johnny Guitar.

Right: Johnny Hutchinson, a temporary Beatle, with the Big Three.

Top: The Cavern, 1961.

Opposite: The Beatles with Ringo in 1963 (top) and bandleader Best with his group in the US some months later (bottom).

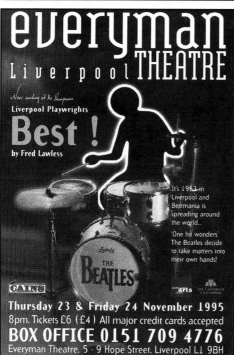

Opposite: Don't look back in anger... Pete Best and mother Mona in 1985.

Above: The Casbah Club, also home to the Best family.

Left: Merseyside hails a local hero.

Following page: Still on the beat at the 1994 Beatles Convention.

Some light on Pete Best's sacking comes from Cynthia Lennon: 'As far as I knew, there was nothing wrong with Pete Best as a musician in any way. He was a very nice fellow, but I think overall, as far as I could see it from my position, their personalities were not in tune with Pete's. Musically yes, but as four personalities they didn't gel. They would have their jokes and Pete wasn't involved in them. He was an outsider. Ringo was nutty enough to fit in with their unit whereas Pete was too serious for them. I think John felt a bit shy and embarrassed about him going, but it's one of those things that happen in life.'

I showed these remarks to Liverpool songwriter Tony Hazzard, who is also a psychotherapist: 'Both George Harrison and Pete Best were quiet, but they were quiet in different ways. Pete was on the outside whereas John, Paul and George could pick up on each other – they shared the same ethos, the same culture, the same sense of humour – and Ringo obviously did as well. There was something about Pete Best that didn't fit. I'd say that Cynthia Lennon had it about right.'

John Lennon told their press officer, Tony Barrow, 'Pete Best was a great drummer, but Ringo was a Beatle.' Tony knew what he meant and adds, 'Ringo had a very dry sense of humour, but he had little to say and in the early days he sat back and let the others do the talking, partly because he was the new boy and the other three knew each other so well over the years. He became known as the Quiet Beatle, the silent one except that when he did come out with something it was pretty original and hilarious.'

Certainly George Harrison preferred to have Ringo in the Beatles because they were such good friends. However, if Pete Best didn't fit in with the Beatles, why did it take them two years of working together to find out?

Alan Clayson, writing in his biography, *Ringo Starr: Straight Man Or Joker?*, explains the sacking thus: 'Pete was a Tony Curtis among the pilzenkopfs, a non-partaker of benzedrine and Preludin, a swain whose intentions towards his Marks and Spencer girlfriend were honourable. Only on the periphery of their private jokes and folklore, and as reliable as he was mature, there was no denying that, to his fellow Beatles, Best was a bit, well…you know. Anyway, he had to go.' This sounds impressive, but why should Pete be derided for having a girlfriend who worked at Marks and Spencer? Is it all that different from Ringo's hairdressing girlfriend?

Compare this assessment with the February 1965 edition of *Playboy* where John Lennon says, 'Ringo used to fill in sometimes if our drummer was ill. With his periodic illness.' and Ringo adds, 'He took little pills to make him ill.' The implication was that Pete Best was a drug addict and Pete obtained damages and an apology.

That assertion was rich coming from John Lennon. Bob Wooler: 'I left the table at the Black Rose one afternoon and when I got back there were two pills floating in my drink. I said, "What's that?" and Lennon said, "Oh, give it here" and he knocked it back. It was two Preludin tablets and they had brought them back from Germany. They were in metal tubes and I used to say to them, "Anyone travelling by tube tonight?"'

THE EX-FILES 7
Pete Refused To Have A Beatle Haircut

We've dealt with this earlier and it would be dismissed for triviality were it not often cited. For example, Paddy Delaney, the doorman of the Cavern, believes it to be a major factor. As stated earlier, Pete's hair was naturally curly and a Beatle haircut would not have been possible. In any event, this is hindsight as who could say in 1962 what a Beatle should look like? Besides, did Charlie Watts ever look like a Rolling Stone?

Surprisingly, Jimmy Nicol, the stand-in drummer for Ringo Starr when he had his tonsils out, said in a 1986 issue of *Drumming* magazine: 'Best was like a cry-baby. He didn't want to cut his hair like the rest of the group and he resented Brian telling him that he had to. He soon found out that Brian carried more weight in the Beatles than he believed. The crap he wrote afterwards about the rest of the band being jealous of his good looks was just wishful babbling. Paul McCartney was ten times the looker Pete Best was.'

Having said all that, the first thing that John Lennon said to Ringo Starr when he became a Beatle was 'Those sidies will have to go.' From a marketing point of view, Eppy was happier with four moptops.

THE EX-FILES 8
Mona Best's Interference

Allan Williams wrote a book, *The Man Who Gave The Beatles Away*, but the title was a nonsense. He had long since relinquished any management role in the Beatles and if anyone could be said to be the Beatles' manager, or at least their agent, before Brian Epstein, then it would be Mona Best. Mona Best. She was The Woman Who Gave The Beatles Away.

Ritchie Galvin of Earl Preston and the TTs: 'Mrs Best was a formidable lady. If she said it was Sunday when it was Tuesday, you'd say it was Sunday too.'

Mona Best was one of the few parents in the early 1960s who understood what Liverpool teenagers wanted. She was very pleasant and very organised, but she could be a harridan with the Best of intentions if she needed to push for the Beatles' interests. To quote Philip Norman in *Shout! The True Story Of The Beatles*, 'Mona Best was a force one did not lightly provoke.'

Despite the fact that she was coping with a pregnancy late in life in 1962, it seems likely that she would have called Brian Epstein regularly with her ideas for the Beatles' progress and advice on what he should do next. A case of mother knows Best. We know from Bobby Graham that Eppy resented calls from Mrs Best as he wanted to manage the Beatles in his own way, which included taking them well away from Liverpool. If Pete Best were no longer with the band, Eppy wouldn't have to justify his decisions to Mona Best.

There is a flaw in this argument. Brian Epstein offered Pete a replacement job as drummer with the Merseybeats and so he would still have to deal with Mona Best. I think he was taking a chance. He guessed that Pete Best would be so upset that he wouldn't want to have anything further to do with NEMS. He primed his friend, Joe Flannery, from another agency so Pete was instead put with Lee Curtis and the All Stars.

THE EX-FILES 9
Pete Best Rejected
Brian Epstein's Advances

True, but Eppy was used to be being turned down. After all, he could hardly expect most of the heterosexuals he approached to leap into bed with him. His advances were also rejected by John Lennon and Billy J Kramer, so it is unlikely to have been a factor.

THE EX-FILES 10
Pete Best Was Unreliable

The first *Anthology* programme in 1996 suggested that Pete Best was unreliable. Pete comments, 'That was the first time I'd heard it. It came as quite a surprise as I wasn't unreliable. I was never criticised for timekeeping as I am a prompt, punctual person. I only missed two gigs with them – one was when I had to go to court and the other was when I had flu. I missed the afternoon session at the Cavern but I pulled myself out of my sickbed to turn up for the evening one, but to my knowledge they were the only two I missed.'

Bob Wooler, DJ at the Cavern: 'It is absolute rubbish to say Pete was unreliable. The most unreliable person was Paul McCartney, who was consistently late. I saw him on TV saying that Stevie Wonder was a bit unreliable, he turned up late, and I thought, "Look who's talking". I would say to Paul at Aintree Institute, "You've missed the middle spot and you'll have to go on last, which is the going home time." He'd say, "Sorry, I was busy writing a song." That didn't impress me at all at the time as I had a show to put on. John, surprisingly, was quite dutiful. Maybe Aunt Mimi was the one behind him, telling him to get out of the house.'

THE EX-FILES 11
It Was Just Hard Luck

Rick Wakeman: 'When you know what the future is, it's easy to criticise the past. Nobody could have foreseen what would happen to the Beatles and Pete was just unlucky. Who knows why some people leave some groups? It can be something that doesn't have a verbal explanation, something intangible, you know. It didn't feel right, and no-one knows why.'

Some years ago I broadcast an interview with Michael Clarke, the former drummer with the Byrds, on BBC Radio Merseyside. When I asked him why he left the Byrds, he answered, 'Because I hated all of them.' Which seemed a good enough reason to me.

The next day I met Joe Butler of the Hillsiders and he said, 'I've never heard anyone say that before and it's so true. When you're seeing so much of each other in a band, likes and dislikes get magnified out of proportion. There's usually someone that everyone hates, then next week, for no reason at all, it's somebody else.'

Could that have happened with the Beatles? Did it just happen to be Pete's week, poor bloke? Probably not as the plot had been simmering for months, but the only person who might defend him was preoccupied and it was hard luck from that point of view.

THE EX-FILES 12
Instant Karma

The interview with George Harrison on the *Anthology 1* video is revealing. When asked about Pete Best, he looks away from the camera and says, 'Historically, it may look like we did something nasty to Pete. It may have been that we could have done it better but the thing was, as history shows, Ringo was a member of the band, it's just that he didn't enter the film until that particular scene.'

This suggestion, treating the Beatles as children of destiny, is typical of Harrison's later, fanciful ideas and I'm sure he never felt that way in 1963. Nevertheless, it is not far removed from the views of the John Lennon biographer Paul Du Noyer: 'Pete Best's sacking is one of the great enigmas of musical history. Was it because he was too good-looking? You can say with hindsight that Ringo Starr fundamentally affected the chemistry of the group. Lots of things about the Beatles seem slightly magical, not quite explicable in ordinary terms, and one of the magical things is the interaction of these four personalities. The presence of Ringo acted as a subtle counterbalance to the personalities of the other three, and it was all part and parcel of what made the Beatles happen in the end. Had it not been Ringo, they might not have taken off to the same extent. They might not have captured the subconsciousness of the population.'

In a similar vein, we have George Melly: 'I think they were right in getting rid of Pete Best and recruiting Ringo Starr. Pete Best was tremendously popular in Liverpool and undoubtedly it was a great tragedy for him to be sacked at the very moment when they were breaking through but whoever it was, be it Brian Epstein, George Martin or the Beatles, whoever it was, saw that Ringo's personality was the perfect foil. He was plain whereas the others were all rather good-looking. He was thick whereas the others were rather bright. He was working class whereas the others were basically suburban. Ringo completed the Beatles and made them much more effective, not just musically but as personalities.' A brilliant summary, but a case of being wise after the event – I'm sure not even Epstein had determined that the Beatles needed a 'plain, thick, working-class' drummer.

THE X FILES 13
Who Knows?

Some reasons may be kept private. I know someone who left a group because he wanted to bed one of his partners' wives while they were touring, and another who was sacked because he was stealing from his partners. I've no grounds for supposing there is anything untoward here, but no-one can ever know for certain.

I LOOK TO FIND A REASON TO BELIEVE

So why did Pete Best get the Big E? We can summarise the reasons as follows:

Reason for sacking	John	Paul	George	Brian Epstein	George Martin
Poor drumming	√	√√√	√		√√
No versatility		√√	√		√√
George Martin disliked drumming	√√√	√√√	√√	√√	
Ringo preferred	√	√√	√√√		
Jealous of good looks	√	√√	√		
Didn't fit in	√	√√√	√√		
No Beatles haircut				√	
Mona Best's interference				√√√	
Rejected Eppy's advances					
Unreliable		(√)	(√)		
Hard luck					
Instant karma			(√)		
Unknown factors					

Everybody wanted Pete Best out of the Beatles but for different reasons.

The prime mover is Paul McCartney. After moving to the bass when Stu Sutcliffe left, Paul realised Pete's limitations. In unguarded moments after the sacking, he told Merseybeat musicians, 'Pete wasn't the best drummer in the world – he wasn't even the best drummer in the Beatles.' He was determined that the Beatles should be as strong a musical unit as possible – hence, his desire to remove Stu Sutcliffe and then Pete Best. He was not particularly close to Pete and things came to a head when Pete's drumming was derided by George Martin: Pete would have to go now. Although he never said it, the boyish-looking McCartney was also the Beatle who was most jealous of Pete's good looks.

George Harrison liked Pete Best but as the months went by, found he would rather have his friend, Ringo Starr, in the group. He was pressing for Ringo once Ringo had depped for Pete in February 1962. He, too, was delighted when George Martin thought Pete was the weak link. The fact that George Harrison got the black eye suggests that he played a significant role.

John Lennon was more devil-may-care when it came to musical ability. He tolerated Stu Sutcliffe's musical limitations longer than he should have done. He got on well with Pete, although they only had girls and rock'n'roll in common. At first, he didn't want to disturb the *status quo* of the band but, when distracted with a pregnant girlfriend and a forthcoming marriage, Paul and George seized the opportunity and decided that it was time for Pete to go. What's more, Pete had to be replaced before that all-important first TV appearance.

John was not the instigator of Pete being sacked. If he was, he wouldn't bother to cover it up and he'd have sacked Pete himself, perhaps bringing things to a head in an argument. The reason he felt embarrassed later is because he hadn't done enough to defend him. As he told Hunter Davies from *The Beatles – The Authorised Biography* (1968): 'We were cowards when we sacked him. We made Brian do it. But if we'd told Pete to his face, that would have been much nastier than getting Brian to do it. It probably would have ended in a fight if we'd told him.'

Brian Epstein was happy to go along with the plan as it gave him a freer hand with managing the group, while George Martin, who disliked Pete's drumming, was surprised they'd taken such drastic action.

Whatever the reasons, Pete Best's sacking had dramatic and unforeseen circumstances.

TOMORROW NEVER KNOWS

'Who'd had a record? Arthur Askey was the last one, I think.'
Ringo Starr in *Anthology 1*, Apple video, 1996

I

The filming of the Beatles' lunchtime appearance at the Cavern on 22 August 1962 was a disaster. It wasn't the Beatles' fault – just that the Granada technicians hadn't worked out the best way to capture sound and vision in a noisy, vision-restricted dive like the Cavern. They recorded the Beatles with their new drummer, Ringo Starr, performing 'Some Other Guy' and 'Kansas City'.

The film was not shown but shelved rather than destroyed. Once the Beatles' became nationally famous, it was dusted off and has been screened at regular intervals ever since. Listen hard and you will hear someone shout 'We want Pete!' – and it wasn't Paul or George.

II

Barron Anthony of the Barron Knights: 'The first time I saw the Beatles, it was a whole new ball game. There were these four blokes who fitted so perfectly – Paul was so economical on bass, and Ringo concentrated on simple rhythm patterns, and was unlike most drummers of the day who were too loud or speeding up, putting fill-ins at the end of each verse. The backing was very economical and all three of them harmonised so beautifully and yet it had an oomph to it that I had never heard before. I was knocked out by them and that was before I'd heard their own songs.'

III

George Martin had sent Brian Epstein an acetate of a song he wanted the Beatles to learn. It was a bright, bouncy song called 'How Do You Do It' and was written by a young Tin Pan Alley songwriter, Mitch Murray. He was hoping for a hit with Garry Mills' 'Save A Dream For Me' and Mark Wynter was about to record 'That Kinda Talk', which would become the B-side of his Top 10 hit, 'Go Away Little Girl'. Mitch says, 'I wrote "How Do You Do It" but Adam Faith's management had not taken it up. The music publisher Dick James had heard "How Do You Do It" along with a comedy number, "The Beetroot Song", and he said, "I think 'The Beetroot Song' will be a very, very big hit." A singer called Johnny Angel was going to record "How Do You Do It" but he changed his mind and recorded another song of mine, "Better Luck

Next Time", so better luck next time, Johnny Angel. The next thing I heard was that a new group from Liverpool was going to record "How Do You Do It" and I said, "I'd prefer a big artist but let's see how it goes."'

IV

Just before the session, Brian Epstein told George Martin that the Beatles had dismissed Pete Best and he had been replaced by Ringo Starr. Epstein told him that he was a better drummer than Pete Best and that the band would prefer to record without a session drummer. George Martin agreed.

So, three months after their Parlophone audition, the Beatles returned to Abbey Road on 4 September to record their first single. They arrived around half-past two to run-through their material with Martin's assistant, Ron Richards. 'How Do You Do It' was a certainty and two of their own songs were chosen, 'Love Me Do' and 'PS I Love You'. They went for a quick meal with George Martin and the session itself was from 7 'til 10pm. They recorded several takes of 'How Do You Do It' and 'Love Me Do' but because of union restrictions, they couldn't work past 10pm and cut the third song.

Mitch Murray: 'The Beatles recorded "How Do You Do It" and I hated it. I felt that something had been screwed up, perhaps deliberately, although it is now very evocative of the early Beatles. I can't blame them because they were songwriters themselves and didn't want to do it, but it was a waste of a good song. I thought it was terrible, and fortunately, Dick James agreed with me. He told George Martin that the Beatles had made a very good demo record. George took it very well and said that he was planning to re-do it with the Beatles later on.'

Paul McCartney had told Martin, 'We can't go back to Liverpool singing "How Do You Do It". We can't be seen with that song.' It was recognised as a commercial song but it was nursery-rhyme pop as far as the Beatles were concerned. As Paul says in *Many Years From Now*, 'We knew that peer pressure back in Liverpool would not allow us to do "How Do You Do It". We knew we couldn't hold our heads up with that sort of rock-a-pop-a-ballad. We would be spurned and cast into the wilderness.'

So, George Martin gave up on 'How Do You Do It' and realised 'Love Me Do' would be a more suitable single, but there was a problem – Ringo's drumming.

V

The recording engineer, Norman 'Hurricane' Smith, told Mark Lewisohn (*The Complete Beatles Recording Sessions*, 1988), 'I've a feeling that Paul wasn't too happy with Ringo's drumming, and felt that it could be better. He didn't make too good a job of it. I remember too that there was a fair bit of editing to be done.'

George Martin, Ron Richards and Norman Smith thought the same as Paul and it was agreed that 'Love Me Do' would be re-recorded a week later with a session drummer. The drummer was 32 year old Andy White who was married to Lyn Cornell, formerly of the Liverpool group the Vernons Girls.

Andy White told the US *Drumming* magazine (1986), 'George called me because I had the reputation for being a rock drummer. I happened to be working on the rock'n'roll shows *Drumbeat* and *Oh Boy!*. I got the reputation of being a rock drummer even though I wasn't really a rock drummer. I played with Eddie Cochran and Gene Vincent.'

Despite the slight to his drumming abilities, Ringo attended the session and played tambourine on the new version of 'Love Me Do' and maracas on 'PS I Love You'. They also attempted 'Please Please Me', but George Martin felt the song could be improved if they upped the tempo and added some harmonies. All five musicians received a session fee of £5.15s (£5.75) – John Lennon's was sent to 251 Mew Love Avenue (*sic*).

Ringo Starr told *Drumming* (1981): 'I'm not sure about this, but one of the reasons they also asked Pete to leave was George Martin didn't like Pete's drumming. When I went down to play, he didn't like me either, so he called on Andy White, a professional session man, to play the session. There were two versions but you can't spot the difference in the drumming because all I did was what he did because that's what they wanted for the song.'

VI

Mitch Murray: 'It makes me cringe to think that George Martin told the Beatles to come up with a song as good as mine, but he knew that I was a professional songwriter and he liked the song. Brian Epstein then suggested that Gerry and the Pacemakers should do the song instead and I was told he was like a Liverpool Bobby Darin. George Martin asked me to hear Gerry at the Cavern but I said, "I don't care what he sounds like live, it's the record that counts." Arrogant little sod, wasn't I? They made the record, I loved it and it delighted me that it got to Number 1.'

Gerry Marsden: 'I thought at the time that the Beatles had just made a demo for us, which was sent to us in Germany. We were very surprised when we recorded it because we had never heard our voices played back, apart from on crummy old tape recorders. We couldn't believe it was us. It sounded really good, but, blooming heck, we never thought it would be a hit.' (Again, this is misleading as, the very week that I am writing this, some tapes of Gerry and the Pacemakers, professionally recorded at Lambda Records in Crosby in 1961, have come to light.)

Gerry's drumming brother, Fred: 'At that stage, we were going to do "Hello Little Girl" but Brian Epstein suggested that we did "How Do You Do It" instead and he would give "Hello Little Girl" to the Fourmost. We heard the Beatles' demo and we decided to put a heavier beat on it than the Beatles. We had been playing together for several years and we really wanted to make a hit record. It didn't matter to us if a song was poppy because having a hit was more important. The Beatles should have stuck with "How Do You Do It" as they might have had their first three records at Number 1. Instead, we're the group who did that.'

Mitch Murray: 'When you have a Number 1, you think, "Phew, at last": it's not bottles of champagne but relief. Then you think "Maybe it's a fluke" and you spend your whole career trying to prove yourself. I wrote it "I Like It" for Gerry's follow-up but John Lennon had given him "Hello Little Girl". John threatened to thump me if I got the follow-up and I thought it was a worth a thump. "I Like It" had the same cheekiness and innuendo and it also went to Number 1. I didn't get a thump.'

VII

On Friday 5 October 1962, the Beatles' first Parlophone single was released – 'Love Me Do'/'PS I Love You'. Much to Ringo's delight, George Martin had chosen the first recording with him on drums.

Many people believe that Brian Epstein bought thousands of copies to hype it into the charts. Firstly, this wouldn't work as the sales charts are compiled from statistics from many different shops across the UK. Secondly, NEMS was the largest record retailer in Liverpool and was always going to sell thousands of copies. Brian Epstein might order 1,000 copies of a potentially big single, and my guess is that he ordered 5,000 'Love Me Do's and secured them on very favourable terms.

According to *The Guinness Book Of British Hit Singles*, 'Love Me Do' made Number 17 on the UK charts, which is following the trade chart in *Record Retailer*. Most music fans went by the chart in *New Musical Express* and there the single made Number 27 for one week on 27 October 1962. Strangely enough, the difference in positions was the other way round for the second single, 'Please Please Me' – the single was Number 2 in *Record Retailer* and Number 1 in the *NME* – which means that the perennial quiz question, 'What was the Beatles' first Number 1?' does not have a definitive answer. NEMS was such a big record retailer that it released its own Top 10 each week for its shop windows and for publication in *Liverpool Echo*. According to the chart, proudly displayed on a cardboard rectangle in their windows, 'Love Me Do' went straight to Number 1.

'Telstar' by the Tornadoes was Number 1 on both charts, and several other classics were doing well – 'The Locomotion' (Little Eva), 'It Might As Well Rain Until September' (Carole King), 'Let's Dance' (Chris Montez), 'Sherry' (Four Seasons), 'I Remember You' (Frank Ifield) and 'The James Bond Theme' (John Barry). Entering at Number 26 was another Liverpool performer, Billy Fury, who was bravely covering an Elvis Presley album track, 'Because Of Love'. The only other Parlophone single on the chart was the country-styled 'Don't That Beat All' by Adam Faith.

An interesting fact which may or may not be of significance: *Record Retailer* published a list of the shops participating in their survey but the *NME* didn't. Did Eppy put any pressure on the named shops to push 'Love Me Do'? If he were a less principled businessman, I'd have said yes, but being Brian Epstein, I doubt it. In all, 'Love Me Do' sold a respectable 100,000 copies, but it deserved to do better. In 1964 it was a US Number 1 and in 1982, a UK Number 4.

Nowadays so many different versions of hit songs are released that it is difficult to determine which is the definitive version. That didn't happen in 1962 as there was usually one only version of a hit song. It is odd, therefore, that George Martin decided to go with the Andy White version on later pressings of the single and on the 'Please Please Me' LP, released in March 1963. To distinguish between them – listen for the tambourine.

VIII

On 12 October 1962, Little Richard came to the Tower Ballroom, New Brighton with a supporting bill that featured so many of the people caught up in this extraordinary drama. It included three people who had drummed with the Beatles that year – Ringo Starr, Pete Best (with Lee Curtis) and Johnny Hutch (with the Big Three). Then there was Ringo's old group, Rory Storm and the Hurricanes, and the Merseybeats, whose drummer, John Banks, could have lost his job to

Pete Best. The promoter, Bob Wooler, must have had fun in putting that show together.

He was also running a book. Rock'n'roll vocalist Karl Terry: 'Bob Wooler was offering bets that the Beatles would be bigger than the Shadows within six months and nobody believed it.' I wonder what the odds were, how the success of the two bands would be measured and, most of all, whether Pete Best had a flutter.

Disk jockey Alan Freeman recalls: 'The Beatles were well known on Merseyside, and we were all waiting for a new musical explosion. The world was ready for a new sound even though it was quite an old sound. It was fresh and very invigorating.'

IX

Cynthia Lennon: 'It was someone from the *Daily Mirror* who finally found out John and I were married. One afternoon I was taking Julian for a walk and this group of photographers had been hanging around for a few days. They followed me around and snatched some photographs. I was trying desperately to keep them away, saying it was my twin sister's baby, but by the end of the day it came out.'

X

Joe Flannery managed his brother, Peter, who became Lee Curtis by reversing the name of the American singer produced by Phil Spector, Curtis Lee. Ironically, Lee Curtis and the All Stars were signed by Decca and Pete Best was featured on their storming single, 'Let's Stomp'. Lee Curtis: '"Let's Stomp" was diabolical. The original by Bobby Comstock was absolutely brilliant, and as the Stomp was popular on Merseyside, we'd asked Decca if we could cover it. They wanted us to rave madly at the end by doing a repeat of the words, "Let's stomp". We were a bit green and I think we repeated the words "Let's stomp" 36 times. I got sick of counting. The record was released and it died.'

Nothing went right for the All Stars' recording career. Lee Curtis intones this litany, 'We asked Decca if we could do "Twist And Shout". They said no and then a few weeks later, it was a hit on Decca for Brian Poole and the Tremeloes. We asked if we could do "Money". They said no and it was a hit on Decca for Bern Elliott and the Fenmen. Next we asked if we could do "Shout" and it was a hit on Decca for Lulu with the Luvvers. We asked if we could do "It's Only Make Believe". They said no and it was a hit on Decca for Billy Fury. We wasted our opportunities with "Let's Stomp" and that monotonous ending.'

XI

Jimmy Tushingham, Ringo Starr's replacement in Rory Storm and the Hurricanes: 'Ringo was the only one of the Beatles who kept in touch. When he made it with the Beatles, he had to declare everything to the Revenue. We got hit for an income tax bill which stemmed from Ringo having to disclose what he had been doing. He sent a cheque to Rory, saying he was sorry about what happened but to put this towards the bill.'

By taking Pete Best's drumseat, Ringo Starr became the luckiest man in the world. He became a star in his own right and he even took lead vocals on the Beatles' records, proving everything sounds good on LSD. Tony Barrow: 'The interest in each individual Beatle seemed to move of its own accord, in natural cycles as it were. From time to time there was a conscious effort on our part to decide, "Well, George hasn't done much in the way of interviews – let's get a couple of George solo interviews." But by and large that worked itself out in the end because the kids themselves would have an "I love Paul" week or month or "I love John" week or month and this was reflected in the fan mail. It did shift around and at one stage the kids did feel that Ringo was being left out of things and sitting at the back as it were. There was a "Let's all love Ringo" campaign, which was slightly before the record "Ringo For President" and all that hysteria.'

XII

After Lee Curtis and the All Stars, Pete Best formed his own band, with Mona doing her best to help him get established. They cut 'I'm Gonna Knock On Your Door' for Decca in June 1964, but most of the records were made for small companies in America. The band recorded 'Boys', 'Rock And Roll Music' and 'Kansas City' and, rather cheekily, Savage Records called an album 'Best Of The Beatles'. Some good, original material was written by band members Wayne Bickerton and Tony Waddington, who went on to have success writing and producing the Rubettes. The Pete Best Band's most significant UK gig was playing at the re-opening of the Cavern in July 1966.

Wayne Bickerton: 'Having an ex-Beatle in the band worked against us. We didn't have the opportunity to develop. It was always "Why did you leave the Beatles?" instead of "What's the future of the band?" The fact that Pete was an ex-Beatle overshadowed everything else, and the talent in the band was a secondary situation. Calling an album "Best Of The Beatles" was ridiculous, to say the least.'

Every newspaper that Pete Best opened told him how the Beatles were becoming more and more successful and indeed, reaching a level of international fame that had never been experienced before. 'I couldn't get away from it. I was aware of it all the time.'

The Beatles never gave him credit in interviews, culminating in the *Playboy* feature in which John and Ringo inferred he was a drug addict. Pete didn't have the finances to take full legal action and had to agree to an out-of-court settlement. The damage had been done as references to Pete as a drug addict appeared in other American magazines. All this contributed to his general malaise.

Pete had married his girlfriend, Kathy, and lived in a first floor flat in the family home at Hayman's Green. In 1965 when Kathy was away visiting her mother, Pete decided to kill himself. He blocked all the gaps in the bedroom, put a pillow in front of the gas fire, turned on the jets and waited for the end. As it happened, Pete hadn't blocked the gaps thoroughly and the gas seeped on to the landing outside. Pete's brother, Rory, smelt it as he was passing, tried the door and broke it down. Mo and Rory coaxed him back to life. 'That was the worst period in my life,' says Pete, 'Nothing like that ever happened again.' Maybe not, but he did have the ignominy of being described as dead in a Trivial Pursuit question.

XIII

In nearly every band that makes it, there is a Pete Best – someone who just missed out on the glories. There is a Pete Best Spice, a Pete Best Rolling Stone, a Pete Best Who, a Pete Best Nirvana and a Pete Best member, or rather non-member, of Oasis. How they adjust to the situation is worthy of a book in itself. Also, I have thought from time to time about writing a book about those on the periphery of the Beatles' scene – Pete Best, Alf Bicknell, Joe Flannery, Sam Leach, Uncle Charlie Lennon, Mike McCartney, Father Tom McKenzie and Allan Williams. In many ways, their struggles are as absorbing as the tales of the Beatles themselves.

While the music world was celebrating the Summer of Love and Timothy Leary was proclaiming, 'Tune in, turn on and drop out', Pete Best was dropping out. He was looking for regular employment to provide for his wife and two daughters. 'I thought it would be dead easy to get a job, but people would see on the application forms that I had been a rock drummer for eight years and they had their doubts. I had to take a job, any job, to prove I could do it, so I worked the first 12 months as a bakehouse labourer. Then in 1968, I went to the unemployment office in Old Swan and they asked me if I was interested in working for the Civil Service. I was taken upstairs to see the manager and she told me to start in two weeks' time. I was with them for over 20 years.'

Pete Best worked for the Government employment scheme, Restart, or its equivalent. When a person who had lost his job came to the Restart offices, he or she would be told, 'Pete Best will see you now.' 'I think it helped them,' said Pete. 'Whatever they'd been through, they knew I'd been through it as well. I'd lost the biggest job in show business, millions of pounds and everything that went with it, and here I was giving advice on how to secure employment.'

Pete Best talking to Peter Trollope, 'May The Best Man Win This Time', in the *Liverpool Echo* on 26 August 1988: 'Life's strange – leaving the Beatles set me on a new life. I now have a wonderful wife, great children and I can walk into my local for a pint whenever I want to. I'm happy – I wonder if they can honestly say that.'

Pete was speaking shortly before he appeared at the Merseybeatle Convention at the Adelphi Hotel and just two weeks before the death of his mother, Mona, from a heart attack. Pete wasn't playing at the Convention, but that would happen for a charity appearance at the Grafton Rooms in 1989. John Banks, the Merseybeats' original drummer, had died and Pete took part in his benefit. How ironic that Pete Best should make his comeback drumming for the Merseybeats, the very job he turned down with Brian Epstein.

Kathy Best encouraged her husband to undertake his own Restart. At first, he worked with members of the Merseybeats and Liverpool Express, notably Billy Kinsley, but then he formed a band with his brother, Roag. In 1996 Pete Best was featured on BBC1's *Animal Hospital* as the band found a live snake in an amp they had hired for a UK tour. Pete Best couldn't escape from snakes.

XIV

To mark the tenth anniversary of John Lennon's assassination in December 1990, Yoko Ono gave support to an all-star charity concert at the Pier Head in Liverpool – 'Imagine...John Lennon'. It was a programme of extremes – Lou Reed chilling out with 'Mother' and Kylie Minogue prancing through 'Help!'. Paul McCartney and Ringo Starr decided not to appear live, while George Harrison declined to appear at all – in retrospect, a wise decision as battles raged over the monies raised by the concert. Paul submitted a video of a medley of 'Love Me Do' and 'PS I Love You', while Ringo evoked the Traveling Wilburys with his video of 'I Call Your Name' with Jeff Lynne and Tom Petty. Despite representations from the Merseycats charity, none of the 1960s Liverpool musicians or bands were allowed on the bill.

Not to be outdone, the Merseycats organised their own charity event, 'Imagine…The Sixties', at the Philharmonic Hall on the previous day. The show was closed by the Pete Best Band featuring Billy Kinsley of the Merseybeats, who included a stomping version of the Beatles' 'I'll Get You'. The compere, Billy Butler, introduced Pete as 'The only Beatle performing in Liverpool this weekend.' Would Pete Best have appeared on a John Lennon tribute if he believed that John was responsible for ousting him from the Beatles?

XV

Pete Best took early retirement in 1993. His brother, Roag, had an entertainment agency, Splash, and played drums for a contemporary band. Roag says 'I went through a heavy metal phase where you play like Animal off the Muppets. You hit everything as hard and fast as you can. Then jazz and funk came along, so what I do is a fusion of everything that has come along since the 1960s.'

They amalgamated their resources and formed the Pete Best Band, which has been touring the UK, the USA, Australia, Japan, Canada, Italy and Argentina. The band plays a mixture of old and new material, but many Beatle covers have been radically changed. It is like hearing the Beatles' hits played by Free. Pete and Roag create a very solid sound with their double-drumming, but surely fans want at least one number where Pete is drumming on his own. The duo can hardly criticise listeners who think that Pete's drumming is being deliberately obscured.

The Pete Best Band consists of young musicians, mostly known to Roag through his other ventures. They have also made several albums together and a surprising assessment came when I played some of their music to drumming authority Garry Tamlyn. 'It sounds like one drummer to me, I can't hear any layering of rhythms there. If they are both playing exactly the same, then it is incredibly accurate as there is no flange effect where one snare hits just after the other. It is incredibly in time – or there is just one drummer playing the basic drumbeat. Maybe one of them is just playing cymbals or hi-hat, but it certainly doesn't sound like two drummers. What they do is nothing like the Allman Brothers' early stuff where one drummer would be laying down a basic drumbeat and the other would be ornamenting around various rhythms, so you can clearly discern that there are two types of drumbeats in the one recording.'

Over the years, Pete Best has been involved in various projects relating to his days with the Beatles. He was an adviser on the film *Birth Of The Beatles* (1979) which was shot in Liverpool and directed by Richard Marquand. Ryan Michael plays Pete Best and the drumming in that film is surprisingly good.

XVI

Who can tell what will happen? Who could tell that Robbie Williams would have more success than Gary Barlow after Take That?

In 1995, I saw a very entertaining, new play at the Liverpool Everyman, *Best!*, by Fred Lawless, which was directed by Paul Codman. 'Love Me Do' makes Number 17 but the Beatles' second single flops, while Pete Best has a Number 1. The play creates an alternative universe in which Pete becomes a major star and good jokes abound – Linda McCartney works at the Stanley Abattoir and Cynthia Lennon runs off with a Japanese businessman. The other Beatles decide to take their revenge. Pete Best, played by Tony O'Keeffe, is portrayed as a total shit who is assassinated in the same way as John Lennon.

XVII

The 1994 film *Backbeat* was about Stuart Sutcliffe and it captured the sheer *joie de vivre* of the Beatles in the early 1960s, particularly in Hamburg. Although Stuart may not have contributed much to the Beatles, he was a very decorative element of their stage show. The Beatles are shown, first and foremost, as John Lennon's band – so much so that the weedily-dressed Paul McCartney is only given one song, 'Twenty Flight Rock'. Even 'Long Tall Sally' is put into John's mouth.

In the film, the foul-mouthed John Lennon, played by Ian Hart, anxious to start a fight, says to Pete Best, 'You don't say much, do you?'

Pete Best (Scot Williams) replies, 'Drummers don't talk, you must have noticed that. Just might as well be deaf and dumb, drummers. I mean, when was the last time you heard a drummer say anything? See? You know why, don't you? I'll tell you why. 'Cause nobody fucking listens.'

The real Pete Best says, 'I didn't meet Scot until after the film had been completed, as the film-makers had told him that it wasn't a good idea to meet me. He said he would have portrayed me differently had he met me, which is a nice compliment. They got the publicity wrong: they marketed it as the story of five lads in Germany, but it was really the love story of Astrid and Stu, with John making the eternal triangle.'

Scot Williams: 'I had to mime playing the drums on the soundtrack. Most of the time it was Dave Grohl of Nirvana, but they used Tony Sheridan's version of "My Bonnie" and that was the one I had difficulty with. I couldn't keep up with Pete's playing.'

XVIII

By 1990 the market was flooded with bootlegs of outtakes and alternative versions from the Beatles' recordings sessions, which were easily obtainable through record fairs. The Beatles obtained no income from this and this, more than anything, prompted them to look at their back catalogue and compile a series of CDs and videos under the blanket title 'Anthology'.

They decided that the interviews for the videos should be restricted, by and large, to themselves, George Martin and Neil Aspinall. Pete Best was not asked for his views. When the double-CD 'Anthology 1' was released, the Beatles couldn't resist a further dig at Pete. A well-known poster advertising 'The Savage Young Beatles' was prominently featured. Pete's face had been torn away and replaced by Ringo's. Okay, Klaus Voormann did the artwork but the Beatles approved it. 'When I saw it, I thought, "That's funny, my head's missing",' says Pete good-naturedly. 'It was out of order but it's their project and it's up to them.' Pete got his revenge because Carlsberg filmed an advert for 'Probably the Pete Best lager in the world' that was shown in the advertising break for the first TV episode of the *Anthology* series.

To be fair, 'In My Life' is used for a montage in the first *Anthology* video. As John Lennon sings, 'Although I'll never lose affection for people and things that went before', we're shown a picture of Pete Best.

Apple had to contact Pete Best because 12 tracks with Pete on drums were to be used on the CD. 'Apple told me that they had decided to use certain tracks with myself on the first "Anthology" CD. I was more excited about the world tour I was planning at the time, and the terms and conditions for the "Anthology" were handled by my lawyers and their lawyers. I stood back in amazement when it was released. There was massive publicity and it was Beatlemania all over again.'

Pete didn't meet Paul, George or Ringo over the 'Anthology' project. It would be difficult for them to meet Pete – it could not be like old friends meeting up because it is tied up with lost opportunities and lost fortunes. 'I could ask Neil Aspinall, who is the head of the Apple Corporation and very busy, to make an appointment with them, but it would be a very false way of renewing a friendship,' Pete says wisely.

'I think something will happen only if it is meant to happen. If I did meet them, I don't think I would ask them why they kicked me out of the Beatles. That would be unfair and not something I would rake up again.'

'Anthology 1' sold for £20 – and even if Pete only received 1p for each track, the album did sell ten million, that's over £1 million. According to press reports, he stood to make £8 million. In any event, it has made him a multi-millionaire. Shortly after its success, I invited Pete Best into BBC Radio Merseyside to answer questions from listeners. The first listener asked him, 'Pete, how are you fixed for a loan?'

XIX

Paul McCartney was knighted in the New Year's Honours for 1997. He is a member of the establishment and yet he has sought to maintain his links with the young. In 1997, for example, he released his symphony, 'Standing Stone', and also attempted to challenge Oasis with a new album. Ironically, the title of his new album, 'Flaming Pie', and the title of the Oasis album, 'Be Here Now', were taken from phrases by John Lennon.

I applaud all that McCartney has done for young musicians, actors and technicians by founding and building the Liverpool Institute of Performing Arts (LIPA), appropriately on the grounds of his old school, the Liverpool Institute. McCartney is as pleasant as someone who is ultra-famous and rich and in charge of a big business empire can be, and he is keen to be seen that way. As. psychotherapist Tony Hazzard says, 'In Jungian terms, you would say that Paul McCartney is all persona. He looks after his image very carefully.'

There are occasional lapses, normally related to his possessiveness about Beatles' memorabilia. He stopped Mal Evans' widow from selling one of his handwritten lyrics and he refused to sign TV host Chris Evans' star-studded *TFI Friday* table because he thought someone would make a great deal of money out of it one day. Although many personal details are given in McCartney's songs, he has never referred to the sacking – or if he has, it has been so oblique, I haven't recognised it.

Paul McCartney was Best Songwriter at the 1997 *Q* Awards, beating such nominees as Paul Weller and Beck. He collected his award but left early when a special award was being given to Phil Spector. Spector had produced, or rather over-produced, the Beatles' album 'Let It Be' and McCartney had never forgiven him for the choir added to 'The Long And Winding Road'. According to *Q* magazine's report, he immediately left the room, muttering 'It's a Let It Be thing.'

XX

In January 1998, the *Liverpool Echo* announced a telephone poll to choose Merseyside's greatest ever entertainers. 100 entertainers were listed with each one allocated a personal phone line to register votes. The motley crew included Arthur Askey, Pete Best, Cilla Black, Sir Adrian Boult, Mel C of the Spice Girls, Lewis Collins, Kenny Everett, Billy Fury, Hetty King, Billy J Kramer, Charlie Landsborough, Freddie Starr, Frankie Vaughan, Bob Wooler and the four moptops. I'm surprised that Red Rum wasn't included.

The results were published in the 23 January edition. Ken Dodd was the winner, so no surprise there. Second came Gladys Ambrose, a semi-operatic singer who plays Julia Brogan in *Brookside*, followed by Sonia, actor Andrew Schofield and Pete Best. How can Pete Best possibly be a greater entertainer than the other Beatles? And for that matter, how can Sonia come third? The result has to be the consequence of multiple voting – Sonia comes from a large family and Pete Best has many supporters. You could argue that Pete Best was getting the sympathy vote, but surely John Lennon, who had been assassinated, deserved even more on that score. However, I'm in no position to carp as I misused my own vote. When I saw my friend, the 1950s pop singer Russ Hamilton ('We Will Make Love', 1957), among the nominees, I voted for him.

The rest of the Top 10 was Billy Fury at sixth, then Keith Chegwin, radio DJ Billy Butler, Rory Storm and at tenth John Lennon. Pete Best Number 5, John Lennon Number 10 and Paul McCartney's position not even published. Strange days indeed.

DISCOGRAPHY 1
The Pete Best Sessions

Up to the 1980s, very few performances of the Beatles with Pete Best were available. The following, with one possible, noted exception, are now known to exist.

FIRST SESSION
Tony Sheridan And The Beatles
With Pete Best

In June 1961, Tony Sheridan and the Beatles (Pete Best, George Harrison, John Lennon and Paul McCartney) recorded at the Friedrich Ebert Halle, Hamburg, that is, not at an official recording studio but on location in a school hall. At first, the records were credited to Tony Sheridan and the Beat Brothers. Produced by Bert Kaempfert, the recordings were:

* My Bonnie
(*Traditional, arranged by Tony Sheridan*)

'My Bonnie' is about a claimant to the throne, Bonnie Prince Charles, who landed in Scotland in 1745 to start a rebellion. His troops were routed at Culloden and he spent 40 years as a fugitive in Europe. Ray Charles gave 'My Bonnie' an R&B treatment in 1958.

Sheridan recorded two versions, one with a slow introduction in German and another with one in English. Often the record is released without the introduction at all. In a new variation, Paul McCartney talks over part of the English introduction on 'Anthology 1'.

* Nobody's Child
(*Cy Coben/Mel Foree*)

In 1948, Mel Foree, a song-plugger for Acuff-Rose, had the unenviable task of trying to stop Hank Williams from drinking. Around that time, he wrote the country weepie, 'Nobody's Child', which was recorded by Hank Snow in 1949. Lonnie Donegan recorded the song in 1956 for his 'Lonnie Donegan Showcase' album. His slow, bluesy treatment was Sheridan's template, although Sheridan omitted his narration.

In the main, Tony Sheridan was a wild rock'n'roller, but John McNally of the Searchers recalls him at the Star-Club: 'Sheridan was always best late at night when he'd got a few drinks inside him. He'd become very melancholy and do the blues. He was the best guitarist around and I'd watch him every night. He did a great version of "Nobody's Child" which was very slow and dynamic.'

Perhaps the Beatles should have brought 'Nobody's Child' back to Liverpool. Lonnie Donegan: 'I find the people in the North of England are much more emotional than the South. I can bring people in Liverpool to tears with "Nobody's Child" and I couldn't do that in London where they are more cynical.'

Although only two verses and choruses, the original recording by Tony Sheridan and the Beatles lasts three minutes 45 seconds, and their version has been truncated on some compilations. Accompanying himself on guitar, Sheridan recorded a six-minute version in 1964. George Harrison encouraged the Traveling Wilburys to record the song for a charity single in 1990.

* The Saints
(Traditional, arranged by Tony Sheridan)

'When The Saints Go Marching In' was a favourite with jazz bands and was taken up by some rock'n'rollers – Bill Haley and his Comets (1955), Fats Domino (1958), Jerry Lee Lewis (1958) and the Isley Brothers (1959). Bill Haley's version is usually quoted as the source for Sheridan's version, which is unlikely as Haley changes the words to showcase his band – eg 'When ol' Rudy begins to play'. Because the Comets' drummer is given a solo, I wish Sheridan had followed Haley's arrangement.

Normally, the song is repetitive with a slight change for each verse – Fats Domino, with a walloping bass drum, sings 'When the Saints go marching in' four times and 'When the sun refuse to shine' once. Jerry Lee Lewis offers a variant with a different melody and gospel-styled verses about Paul and Silas. Sheridan sings 'When the Saints go marching in' three times, 'When the sun begin (not 'refuse') to shine' twice and 'When the ol' Lord calls me there' once, and this places his version closer to the Isley Brothers'.

One of Sheridan's early groups was called the Saints and so he was very familiar with the work. On the recording, Tony Sheridan sounds very like Conway Twitty, but this impersonation is of his own doing as Twitty never recorded 'The Saints'.

In 1963 the Searchers revived 'The Saints' as 'Saints And Searchers' for the B-side of their second hit, 'Sugar And Spice'.

* Take Out Some Insurance On Me Baby

(*Sometimes shown as Charles Singleton/Waldenese Hall, but more likely Jesse Stone, who was also known as Charles Calhoun. Confused? You should be.*)

The blues guitarist Jimmy Reed introduced many songs into the beat groups' repertoires: 'Ain't That Lovin' You Baby', 'Bright Lights Big City', 'Baby What You Want Me To Do' and 'Big Boss Man'. He recorded 'Take Out Some Insurance On Me Baby', which is also known as 'If You Love Me Baby' after its first line, in Chicago in March 1959.

Tony Sheridan gives a creditable performance but his reference to 'goddamned insurance' has been removed on a doctored version, which adds guitar, drums and harmonica.

* Why (Can't You Love Me Again)

(*Tony Sheridan/Bill Crompton*)

Tony Sheridan wrote this song with the British rock'n'roll performer Bill Crompton in 1958. Crompton recorded for Fontana and had considerable airplay on 'A Hoot And A Holler'. Crompton also wrote a Top 20 instrumental, 'The Stranger', for the Shadows. The clumsily-titled 'Why (Can't You Love Me Again)' is an echo-drenched doowop ballad with a forceful middle-eight that sounds as though it belongs in another song.

Gerry Marsden: 'I thought "Why" was a nice song and I wanted to record it, but Tony said, "Hang on, I've got a better one" and gave me "Please Let Them Be", which I recorded some years later. It was one of the biggest flops in the history of records, but it was a lovely song.'

THE BEATLES WITH PETE BEST

Line-up as previous, but without Tony Sheridan.

* Ain't She Sweet
(*Jack Yellen/Milton Ager*)

Jack Yellen and Milton Ager were a successful pre-war songwriting partnership, their most enduring songs being 'Happy Days Are Here Again' and 'Ain't She Sweet'. The lyricist Jack Yellen had emigrated from Poland and this 1927 success was daring for its time as it used American slang ('ain't') in its title. Umpteen recordings of the song have been made with no one version predominating.

'Ain't She Sweet' was a standard by the time Gene Vincent recorded it in 1956. Effectively, the Beatles followed Vincent's direction, but upped the tempo and omitted the whistling. Possibly Bert Kaempfert advised against whistling as Germans considered it insulting.

There is also a 1959 single by UK rock'n'roller Duffy Power. He says, 'Larry Parnes heard the Gene Vincent version and asked me to do it with Ken Jones' Palais de Dance orchestra. They were good musicians but they knew nothing about rock'n'roll.'

John Lennon took the lead vocal on the Beatles' version, which has been overdubbed (with additional drums!) on some releases. Duffy Power: 'It sounds to me like the Beatles hadn't worked anything out. They were just chugging through the chords and adding a Chuck Berry blues walk. Nice guitar solo though.'

* Cry For A Shadow
(*George Harrison/John Lennon*)

While the Beatles and Rory Storm and the Hurricanes were in Hamburg, the Shadows made the UK charts with the theme from the film *The Frightened City*. Rory was wondering how it went and George Harrison and John Lennon told him they had heard it and improvised the instrumental which became, with gentle humour, 'Cry For A Shadow'. For some time, Rory believed that their instrumental was the Shadows' hit single.

Geoff Taggart, a St Helens musician who wrote 'Breakthru' on the best-selling album 'The Sound Of The Shadows', says, 'This is a riff in E with a chord sequence over the top rather than a melody, and as such it is more reminiscent of the fillers on the Ventures' albums than the Shadows. It starts like the Shadows' "Man Of Mystery" but it's really a quicker version of the John Barry Seven's "Rodeo", which came out in 1958. The George Harrison solo is a dry as a plank, which is more like Joe Brown than Hank Marvin. Indeed, Joe gave Hank his echo unit because he couldn't come to grips with it and Hank used it on "Apache". The chord sequence is the same as Joe Brown's "Shine" and, incidentally, Bruce Channel's "Hey! Baby", although that wasn't released until 1962. And there's nothing wrong with Pete's drumming on the track – he's a good drummer.'

SECOND SESSION
The Beatles

New Year's Day, January 1962, Decca Studios, West Hampstead, London. The producer was Mike Smith who made hit records with Billy Fury, Georgie Fame and the group Dick Rowe signed instead of the Beatles, Brian Poole and the Tremeloes.

Irrespective of the standard of the performances, this is a representative cross-section of their repertoire at the time. There are three oldies, a show song, three originals and eight rock'n'roll covers. The talents represented remained, by and large, the Beatles' favourites – Chuck Berry, the Coasters (twice), Goffin/King, Buddy Holly, Carl Perkins, Phil Spector and Tamla-Motown.

Pete Best: 'Decca was a major label, the company to be with, and we thought hard about the material we were going to play at the audition. It was a good cross-section of numbers, and we recorded them like a live set with just one or two takes on each number. We were trying to be cool, calm and collected about it but there were frogs in our our throats. We weren't on form and we could have been a lot better.'

The Beatles' performances are often described as uninspired, but bear in mind:

1   They had travelled overnight for ten hours in a van with their equipment in a snowstorm.
2   They were travelling back the same way.
3   They were miffed that Brian Epstein had come to London by train and had stayed overnight with relations.
4   They had missed out on New Year's Eve festivities.
5   Mike Smith arrived late because he had been to a party.
6   This was a recording test, never intended for public release.
7   They were cutting 15 songs in an hour.

However, I can't accept Mark Lewisohn's assertion that they were 'nervous' and 'ill at ease' (*The Complete Beatles Chronicle*, Octopus, 1992). The Beatles were familiar with making records through their work with Tony Sheridan, they had been in the German charts, they had previously met Mike Smith at the Cavern and they had played in far more ominous places than this. I think their performances were uninspired (though they have their moments) because they were thoroughly pissed off and their tiredness dulled their voices.

The 15 songs include seven lead vocals from Paul and four each from George and John. It is odd that George was singing as much as John, but perhaps not surprising that Pete didn't do his party piece. What did Mike Smith and his boss, Dick Rowe, miss? Well, the three original songs became chart hits (one for a Decca group), two songs were featured on 'With The Beatles' and two became hits for Decca artists.

Mike Smith says, 'I thought the Beatles were absolutely wonderful on stage and I should have trusted my instincts. They weren't very good in the studio, and really we got to the Beatles too early. Nothing against Pete Best, but Ringo wasn't in the band and they hadn't developed their songwriting. Had I picked up on them six months later, there was no way I couldn't have recognised the quality of their songs. I went with Brian Poole and the Tremeloes because they had been the better band in the studio. So much in this industry depends upon being in the right place at the right time and whether I did the right thing or not I'll never know. In fairness, I don't think I could have worked with them the way that George Martin did – I would have got involved in their bad parts and not encouraged the good ones. When I met them later on, they gave me a two-fingered salute.'

* Besame Mucho
(*Consuelo Velazquez/Sunny Skylar*)

The opera *Goyescas* was written in Paris in 1914 by the Spanish composer, Enrique Granados. Because of the Great War, the opera was premiered in New York in 1916. After the performance, its cast and composer returned by ship to Europe, but it was torpedoed by the Germans and no-one survived. I wonder if the film director of *Titanic* James Cameron knows this story. 'The Nightingale Aria' from the opera was luckier: two pop writers converted it to 'Besame Mucho'.

'Besame Mucho' became the 'La Bamba' of the 1940s. It was a million-selling record for Jimmy Dorsey and his Orchestra with vocalists Kitty Kallen and Bob Eberle in 1943 and the song is featured in the 1944 film musical *Follow The Boys*. The Coasters gave the song a straight treatment in 1960, unfortunately allowing their bass singer to take the lead vocal. Pete Best: 'Doing "Besame Mucho" was Paul's idea. He may have been influenced by the Coasters but it was very much our arrangement, and I like it very much.'

Paul McCartney in *Many Years From Now*: 'It's a minor (key) song and it changes to a major, and where it changes to a major is such a big moment musically. That major change attracted me so much.'

The Beatles often performed 'Besame Mucho', returning to it during the filming of Let It Be. The Decca recording is hurried and was slowed down by George Martin for inclusion on 'Anthology 1'. Despite the 'cha-cha-boom', it is a novelty that doesn't quite come off. Often they performed it on stage as a nod to a Liverpool impresario, 'Besammy Leacho', which does amuse me.

In April 1962 Jet Harris, formerly of the Shadows, took a throbbing instrumental treatment of the song into the Top 30 for Decca. Celebrations took place in Guadalajara in August 1962 to celebrate the song's 20th anniversary. This is where Consuelo Velazquez lived, and many artists who had recorded the song took part in the festivities. No one knew then that the Beatles had also recorded it.

* Crying Waiting Hoping
(*Buddy Holly*)

The British singer-songwriter Harvey Andrews says: 'All the great songwriters of my generation came from Buddy Holly. When rock'n'roll started, there was something about Holly that got to us. He was the first singer-songwriter, although we didn't know it at the time. We don't go back to Little Richard, we don't go back to Fats Domino, we don't go back to Elvis, though we liked all of them. Buddy Holly was the one that the young songwriters could relate to.'

The Quarry Men recorded 'That'll Be The Day' at their first recording session, the Beatles recorded 'Words Of Love' on the 'Beatles For Sale' album, and Paul McCartney liked Holly's songs so much that he bought the catalogue. What, incidentally, will happen to the rights of those songs on the 50th anniversary of Holly's death in 2009? Under normal circumstances, they would fall into public domain, but I anticipate that McCartney's lawyers will work out an extension.

Shortly before his fatal tour, Buddy Holly recorded six new songs in his New York apartment – 'Crying Waiting Hoping', 'Learning The Game', 'Peggy Sue Got Married', 'That Makes It Tough', 'That's What

They Say' and 'What To Do'. They have been released with overdubbed instruments but the unadorned tapes remain the best, largely because of the intimate nature of the songs.

George Harrison sang 'Crying Waiting Hoping' with vocal backup from John and Paul and it is one of the most successful of the Decca tracks.

* Hello Little Girl
(*John Lennon/Paul McCartney*)

John Lennon wrote 'Hello Little Girl' around 1958 and it was one of the first original compositions that the Beatles included in their performances. They had tired of the song by the time they recorded for Parlophone, and Brian Epstein offered it to his other signings. Gerry Marsden turned it down as he wanted to be seen as independent from the Beatles although his run-through is included on 'Gerry And The Pacemakers – The Best Of The EMI Years' (1992). The Fourmost had no such reservations and took the song into the Top 10 in 1963.

* Like Dreamers Do
(*John Lennon/Paul McCartney*)

The influence of 'Besame Mucho' can be heard in this early Paul McCartney composition. Again, it was one of the first original compositions to be included in Beatles' performance. It's a naive song – 'I've waited for your kiss, waited for your bliss' is banal. Nevertheless, the song was passed to the Applejacks who took it into the Top 20 in 1964. As luck would have it, the Applejacks recorded for Decca and were produced by Mike Smith.

* Love Of The Loved
(*John Lennon/Paul McCartney*)

'Love Of The Loved' is another Paul McCartney composition and one he considered passing to Beryl Marsden. Epstein insisted on giving it to his protégé, Cilla Black, and it was her first single. Says Cilla, 'Paul McCartney wrote it and I'd heard the Beatles do it many times in the Cavern. I wanted to do a group arrangement and I was ever so disappointed when I got to the studio and there was brass and everything. I thought it was very jazzy and I didn't think it would be a hit. I preferred the B-side, which Bobby (Willis, her future husband) wrote, "Shy Of Love".' 'Love Of The Loved' made the *NME* Top 30 in October 1963.

* Memphis Tennessee
(*Chuck Berry*)

In *Chuck Berry – The Autobiography* (Faber and Faber, 1987), the rock'n'roller explains that he developed 'Memphis, Tennessee' from a line in a Muddy Waters blues where he was talking to a long distance operator. He writes, 'My wife had relatives there but, other than a couple of concerts there, I never had any basis for choosing Memphis as the location for the story.'

The song about Chuck Berry trying to establish contact with his six year old daughter was released in 1959. It was not a UK hit – indeed, Chuck Berry, had only had one Top 20 hit prior to 1962 – but it was recorded by the Beatles with John Lennon's vocal. The British beat boom sparked an interest in Chuck Berry's work and he had a Top 10 hit with 'Memphis, Tennessee' in late 1963. Unfortunately for him, he was then in prison for corrupting a minor.

* Money (That's What I Want)
(*Berry Gordy Jr/Janie Bradford*)

At first, Berry Gordy Jr, the founder of Tamla-Motown, had success as a songwriter, writing 'Reet Petite' for Jackie Wilson and 'You Got What It Takes' (Marv Johnson). Gordy is quoted in *Motown – The History* (Guinness, 1988) as saying, 'When people asked me what I did for a living, I would say, "I write songs." They would have sons and daughters becoming doctors and lawyers, and my mother and father were embarrassed. Even though I had many hits, I didn't have any money. I came from a business family – my mother and father always talked about the bottom line, and the bottom line is profit.'

'Money (That's What I Want)' was recorded by Barrett Strong for Gordy's new Anna label in 1959 and made the US Top 30. Although it did not chart here, it became a staple song for beat groups. The Beatles, the Searchers, the Undertakers and Kingsize Taylor and the Dominoes recorded Liverpool versions and other covers came from Bern Elliott and the Fenmen and the Rolling Stones, who both recorded for Decca. Elliott made the UK Top 20 around the time that the Beatles' version was released on the LP, 'With The Beatles'. George Harrison, the most-money conscious Beatle, wrote a companion song in 'Taxman'.

The balance on the Decca audition tape puts Pete Best's drums in the background, but it is easy to imagine how they would be thundering out in the Cavern. The various versions enable us to hear the drumming techniques of Pete Best, Ringo Starr and Charlie Watts on the same song. Garry Tamlyn: 'There's a similarity in drumbeat among them all. At the beginning of the verse, we have emphasis on either tom-toms or snare drum in quaver patterns but by the time you hit the chorus section, we have a standard rock beat. There are some differences in rhythm, Pete Best was performing the typical late 1950s, early rock umpapapa, and in the beginning of the verse, he was using tom-toms in semi-quaver rhythm – it is a bit like Jerry Allison's drumming in Buddy Holly's "Peggy Sue". Ringo didn't adopt that semi-quaver tom-tom rhythm, he was playing quaver rhythms on snare drum, maybe with some tom-tom activity. Charlie Watts employed a standard rock beat, backbeat on beats two and four, even quavers, but he did have some tom-tom activity in the verse section. Very much like Ringo Starr, and all of them are very good and competent.'

* Searchin'
(Jerry Leiber/Mike Stoller)

In 1957 the Coasters recorded 'Searchin'' in ten minutes towards the end of a session. Their producers and songwriters, Jerry Leiber and Mike Stoller, discuss the song in the CD booklet for '50 Coastin' Classics' (Rhino, 1992). Stoller says, 'I had worked out an old-timey piano lick that struck me as being kind of fun, and it worked.' Leiber adds, 'Everybody was together. It was one of those moments that rarely happens, and it turned out to be their biggest hit.' Although 'Searchin'' made Number 3 in the US, it was only in the UK Top 30 for one week. The B-side was another rock classic and Beatles' favourite, 'Young Blood'.

The Beatles were performing this tribute to fictional detectives as early as 1958 with Paul McCartney on lead vocals. The Manchester group, the Hollies, were strongly influenced by the Beatles and their first two hits were with Coasters' songs – '(Ain't That) Just Like Me' and 'Searchin''.

In 1982 Paul McCartney chose the Coasters' 'Searchin'' as one of his Desert Island Discs.

* September In The Rain
(Al Dubin/Harry Warren)

Harry Warren was an award-winning songwriter who once told the equally impressive Harold Arlen, 'You walk two Oscars behind me.' Many of his best-known songs were written in Hollywood with the lyricist Al Dubin and include 'I'll String Along With You', 'I Only Have

Eyes For You' and 'Tiptoe Through The Tulips'. 'September In The Rain' was written for the 1937 film musical *Melody For Two*. When the tempestuous jazz singer Dinah Washington revived the song in 1961, it made the UK Top 50 and was a success at the time of the Beatles' recording.

It is mooted that the Beatles held themselves in check at the Decca audition and 'September In The Rain' is a good example. As the song draws to a conclusion, you expect Paul McCartney to go for a big ending but he decides otherwise and the final notes are ragged.

* The Sheik Of Araby
(Harry B Smith/Francis Wheeler/Ted Snyder)

In 1921 Rudolph Valentino starred in *The Sheik* and a ragtime number was written in his honour, 'The Sheik Of Araby'. This, in turn, was used to accompany his silent films. It was featured in a 1940 film musical, *Tin Pan Alley*.

Many regard the Beatles' treatment of 'The Sheik Of Araby' with George Harrison's lead vocal as their worst performance and it was singled out by critics hostile to the 'Anthology' concept. Although hard to defend, it is not without interest. To me, it sounds like an attempt to emulate Joe Brown performing pub favourites like 'Darktown Strutters Ball'. Joe Brown did include 'The Sheik Of Araby' in his stage-act and on his 1963 live album, but his version is more relaxed.

Pete Best says, '"The Sheik Of Araby" was a very popular number and we nearly did it on the BBC shows because of the demand. George loved those kind of numbers and we got it from Joe Brown and the Bruvvers. We rocked it up a bit.'

* Sure To Fall
(Carl Perkins/Bill Cantrell/Quinton Claunch)

The rockabilly singer Carl Perkins, when asked what he thought of the Beatles performing his songs 'Matchbox', 'Honey Don't' and 'Everybody's Trying To Be My Baby', replied, 'When I look at the Beatles, I see great big dollar signs.' He maintained a long friendship with each Beatle and his 1996 album, 'Go Cat Go!', included duets with Paul, George, Ringo and, through modern recording technology, John.

Several Merseyside guitarists (Billy Hatton of the Fourmost, Kingsize Taylor) owned Carl Perkins' 1959 'Dance Album' and one track was the slow country song, 'Sure To Fall'. The lead vocal is taken by Carl's brother, Jay, with Carl singing harmony and playing a very distinctive guitar solo. The Beatles' version is livelier and as close to country harmonies as they ever got.

* Take Good Care Of My Baby
(Gerry Goffin/Carole King)

Diana Mothershaw, who worked at Rushworth's, remembers Paul McCartney coming into the store regularly in 1962: 'He loved Carole King's "It Might As Well Rain Until September" and he would go on about Goffin and King. He and John wanted to be songwriters like that.' The 1997 film *Grace Of My Heart* presented a thinly disguised version of the Goffin/King story and showed how Goffin always wanted to write something more substantial than teenage pop songs. But, as Bobby Vee says, 'Those Brill Building songs have endured. We used to think of them as simple songs but they are *profoundly* simple.'

'Take Good Care Of My Baby' was a transatlantic Number 1 for Bobby Vee in 1961. He remembers, 'The first song of theirs that I recorded was "How Many Tears" and they flew to Los Angeles to be at the session. During one of the breaks, Carole sat down at the piano and sang "Take Good Care Of My Baby" and it was like, Wow, I couldn't believe what I was hearing. Dion had already done a version of it, but my producer Snuffy Garrett had asked her to write an intro and she had come to California to present it in its finished form.'

The Beatles, with George on lead vocal, recorded the song with the intro. Pete Best: 'George tried a couple of Bobby Vee numbers, and he, more than the others, thought we should do one or two from the Top 20.'

* Three Cool Cats
(Jerry Leiber/Mike Stoller)

'Three Cool Cats' was the B-side of the Coasters' transatlantic hit 'Charlie Brown' from 1959. In the CD booklet for '50 Coastin' Classics' (Rhino, 1992), Jerry Leiber says, 'We tried to make a kind of Afro-Cuban sound that Mike used to dig a lot in LA in the early 1950s – but fitting into a Coasters-type format with a funny setting and all of that.'

The song had been performed memorably on ITV's *Oh Boy!* by Cliff Richard, Marty Wilde and Dickie Pride. An intriguing feature of the Beatles' good-humoured version is their cod-Indian accent on one line, which would not be PC today. Mike Smith: 'They did black R&B

songs like "Three Cool Cats" and it wasn't very good. They were overawed by the situation and their personalities didn't come across.'

* Till There Was You
(Meredith Willson)

Fifteen songs were taped at the Decca auditions and only two were subsequently recorded and released by Parlophone. One, inevitably, was 'Money' but the other, quite surprisingly, was the show song 'Till There Was You'.

The Music Man was a 1957 musical about a conman who persuades small town locals to buy band instruments and uniforms but, after falling in love with the librarian, he settles down and becomes the bandleader. Its best-known songs are 'Seventy-six Trombones' and 'Till There Was You', the latter being a Top 30 hit for Peggy Lee in 1961. Paul McCartney was the Beatle most enamoured by musical theatre and film, although, strange to report, he has yet to write a musical himself. He did, however, write and produce a song, 'Let's Love', for Peggy Lee in 1974.

Paul McCartney sounds unsteady and the Beatles play uncertain notes, but this was a sophisticated song being recorded on a one-take audition. George Martin reworked the song with acoustic guitars and put Ringo on bongos, but the 'Anthology' version features Ringo on drums. I asked Garry Tamlyn to compare all three: 'There's very little drum activity from Pete Best, just even quavers on the hi-hat. There are some tempo fluctuations, he pushes the beat a little bit and then pulls it back and so he gets a little out of ensemble with the rest of the band, but he had no tom-tom or snare-drum activity, like Ringo Starr's version. It sounds like Ringo was implementing the same beat as he performed on his bongo-drum recording, and that is very good drumming, quite creative, with lots of responses to the phrasing, and very accurate too.'

* To Know Her Is To Love Her
(Phil Spector)

When Phil Spector attended his father's funeral, the tombstone read, 'To know him is to love him' and, almost immediately, a song was born. He recorded the song in 1958 with his friends Annette Kleinbard and Marshall Leib as the Teddy Bears, and with Sandy Nelson on drums. It was a US Number 1 and a UK Number 2.

The song's construction is similar to Buddy Holly's, so it is easy to see how it appealed to the Beatles. John took the lead vocal, now calling the song 'To Know Her Is To Love Her', and, again in the heat of the moment, he takes the song too fast – the *Live At The BBC* version is superior. In 1965 the song, as 'To Know You Is To Love You', was a Top 10 hit for Peter and Gordon. Another variant would be 'To Know Me Is To Love Me' – perhaps Liam Gallagher could record that.

THIRD SESSION
The Beatles

7 March 1962. Playhouse Theatre, Manchester.
BBC Light Programme *Teenager's Turn*.
This was the Beatles' first radio appearance for which they were paid a total of £30.

* Dream Baby
(*Cindy Walker*)

Cindy Walker, the writer of such country hits as 'When My Blue Moon Turns To Gold Again', 'You Don't Know Me' and 'Distant Drums', met record producer Fred Foster and was asked to submit some songs for Roy Orbison. 'I came home and I wrote "Dream Baby", "Love Star" and "Shahdoroba",' she told Craig Baguley in *Country Music People* (November 1997), 'and I put them in a little box to mail to Fred Foster. Then I took out "Dream Baby" as I thought it was too repetitious, the same thing over and over. My mama said, "If you don't send that, you haven't got anything at all to send", so I put it back in the box. About a week later, Fred Foster called me and played "Dream Baby" by Roy Orbison. It sounded wonderful, and if it hadn't been for my mama I wouldn't have sent it.'

'Dream Baby' was released in February 1962 and entered the UK charts on 8 March 1962, climbing to Number 2. At the BBC audition, the producer Peter Pilbeam wrote 'John Lennon, yes: Paul McCartney, no', but Paul was still allowed to perform the song he had just learnt on their radio debut. Within 15 months, Roy Orbison would be conceding top billing to the Beatles.

* Memphis Tennessee
(*Chuck Berry*)

As before.

* Please Mr Postman

(*Brian Holland/Robert Bateman/Berry Gordy Jr/William Garrett/
Georgia Dobbins/F Gorman or Garman/Brainbert*)

If you have 'Please Mr Postman' in your collection, check out the writers. Like 'Why Do Fools Fall In Love' and 'Louie, Louie', it is a song whose authorship is in doubt. I have listed the seven names that I have found on various versions, but perhaps Gladys Horton, the lead singer of the Marvelettes, has the true story. 'The original "Please Mr Postman" was a blues written by William Garrett,' she told me, 'It was something like (sings) "Please please please I want a letter from my baby, 'Cause she gone and left me." One of the girls in the group, Georgia Dobbins, who was a very good writer, asked William if she could use the title but change the song as we were too young to be singing the blues. BB King could have done the original song, but not us. William had heard that Berry Gordy had picked our group to record at Motown and he said, "You can do anything with the song, but make sure I get a writer's credit." She reconstructed the song and made it about a girl waiting for a letter from her boyfriend, and the only two names that should have been on "Please Mr Postman" were William Garrett and Georgia Dobbins. We then took it to Motown.'

Somewhere along the line, the label's owner, Berry Gordy Jr, and the group's producers, Brian Holland and Robert Bateman, were added to the credits, along with two names I can't identify. 'It happened a lot at Motown,' says Gladys. 'I was the only person who wrote "Playboy" but the producers' names were added to the credits. Why should I have to share the credits?' 'Playboy' was a US Top 10 hit and the credits are particularly significant with 'Please Mr Postman' as the song has become a much-performed rock'n'roll classic.

In December 1961, the Marvelettes' 'Please Mr Postman' was the first Tamla-Motown single to top the US charts, although it failed to make the UK Top 30. 'Sometimes you don't pay attention to the mail 'til someone you love moves away from you, then you're looking for the postman and you think of the Marvelettes,' says Gladys and I have another ten minutes of her praising the US postal system. If she is not paid to publicise their services, she should be.

John Lennon took the lead vocal on the Beatles' version and the song was included on their second album 'With The Beatles'.

FOURTH SESSION
Tony Sheridan And The Beatles

In May 1962, Tony Sheridan, Roy Young (piano) and the Beatles (Pete Best, George Harrison, John Lennon and Paul McCartney) recorded at Studio Rahistedt (also known as Studio Wandsbek), Hamburg and were produced by Bert Kaempfert and his assistant, Paul Murphy. According to Pete Best's autobiography *Beatle!* he recorded 'Sweet Georgia Brown' and 'Skinny Minnie' at this session, but no tapes of the latter exist and it is more likely that Pete has confused it with a later recording of Sheridan's.

* Swanee River
(*Stephen Foster*)

Writing for minstrel shows, Stephen Foster (1826-64) became the first major pop songwriter and his compositions include 'The Old Folks At Home', 'Camptown Races', 'Beautiful Dreamer', 'Jeannie With The Light Brown Hair' and 'Swanee River'. Ray Charles updated the last-named as 'Swanee River Rock' in 1957.

Again, Sheridan recorded the song without the Beatles in December 1961, and the later version with the Beatles has a slow introduction, which is sometimes omitted on reissue. The first version has not been released.

* Sweet Georgia Brown
(*Ben Bernie/Maceo Pinkard/Kenneth Casey*)

This song from the 1920s was associated with Pearl Bailey and Bing Crosby, but had been revived by the Coasters in 1957. Their stop-start arrangement is quite different from Sheridan's.

Sheridan had recorded the song without the Beatles in December 1961, but this version lacks vocal harmonies and lasts 2.25. The version with the Beatles has vocal harmonies and lasts 2.00. In 1964 the Beatles' version was remixed and Sheridan added a new verse, written by Liverpool musician Paul Murphy, about the Beatles' fame – and hairstyle. 'We didn't take that seriously,' says Sheridan. 'I just thought the words were witty and I sang them.'

LIVE RECORDING
The Beatles

A recording made by a Beatles' fan at the Cavern around June 1962 was bought by Paul McCartney at Sotheby's in 1985. He only paid £2,310 for the tape and bootleg versions do not exist.

The Beatles' set list was 'Words Of Love', 'What's Your Name', 'Roll Over Beethoven', 'Ask Me Why', 'Hippy Hippy Shake', 'Till There Was You', 'If You Gotta Make A Fool Of Somebody', 'Please Mr Postman', 'Sharing You', 'Your Feet's Too Big', 'Dizzy Miss Lizzy', 'I Forgot To Remember To Forget', 'Matchbox', 'I Wish I Could Shimmy Like My Sister Kate', 'Memphis, Tennessee', 'Young Blood' and 'Dream Baby'.

FIFTH SESSION
The Beatles

6 June 1962, EMI, Abbey Road, London.

* Besame Mucho
(Consuelo Velazquez/Sunny Skylar)

Another version of the song recorded at the Decca auditions. Again, it is taken too fast, but this was the way the Beatles performed it. It was a strange choice as Jet Harris' version was still in the charts. Garry Tamlyn: 'The snare rhythm that Pete plays is similar to "Love Me Do" and it is very tight drumming, which goes against my comments on "Love Me Do". Maybe he just didn't have enough time rehearsing that song.'

* Love Me Do
(John Lennon/Paul McCartney)

Garry Tamlyn: 'This is slower in tempo than the Ringo Starr and Andy White versions. The cymbals are only audible in the breaks and the middle eights and it is really interesting to hear what Pete Best does. The first time he plays a shuffle rhythm all the way through, and the second time he reverts to an even quaver rhythm on cymbals with a syncopated snare drum rhythm – a Latin beat, really. He has some tempo variations, particularly in the second break. Even his quaver rhythms on cymbals are uneven. On this showing, it doesn't surprise me that George Martin was critical of his drumming.'

Pete Best: 'I did what I thought was right for the number at the time. The idea was to make the middle eight different from the rest of the tune, which is something we had tried and tested in Germany. It is slightly slower and different from the other versions, but we were going to put some finishing touches on it.'

I also played Garry the Ringo Starr and Andy White versions. 'You can't hear the bass drum clearly on Ringo's recording but the snare is very audible. It sounds as though he is playing a swing rhythm on the hi-hat or cymbals. The bass drum is very audible on the version with the session drummer and aligns with the bass guitar riff very well. The session drumming is very tight and very accurate, the best of the bunch.'

* PS I Love You
(*John Lennon/Paul McCartney*)

EMI have not confirmed that this version of 'PS I Love You' still exists.

SIXTH SESSION
The Beatles

11 June 1962. Playhouse Theatre, Manchester.
BBC Light Programme *Here We Go*.

* Ask Me Why
(*John Lennon/Paul McCartney*)

With 'Ask Me Why', John and Paul were trying to write in the style of Motown, albeit with a slight Latin-American influence. It is a fairly simple song but Lennon takes it for all it is worth with the way he stretches and bends words. It became the B-side of 'Please Please Me'. Possibly John wrote it for Cynthia, although 'You're the only one that I've ever had' hardly fits this interpretation.

* Besame Mucho
(*Consuelo Velazquez/Sunny Skylar*)

And again...

* A Picture Of You
(*John Beveridge/Pete Oakman*)

After a succession of steady sellers, Joe Brown topped the *NME* charts in July 1962 with the country-styled 'A Picture Of You', written by two of his Bruvvers. At the time of the Beatles' BBC recording, the song had just entered the Top 10.

Encouraged by George Harrison, the Beatles performed Joe's hits 'Darktown Strutters Ball', 'What A Crazy World We're Living In' and 'A Picture Of You'. Geoff Taggart of St Helens' group the Zephyrs says, 'George was heavily influenced by Joe Brown, who, in turn, had taken so much from Carl Perkins and Paul Burlison of the Johnny Burnette Trio. Often I hear George Harrison solos on Beatle records and I think, "He's doing Joe Brown again."' George now lives near Joe in Henley-on-Thames and they have had private sessions recording George Formby songs for their own amusement.

DISCOGRAPHY 2
Pete Best And Ringo Starr

If you want to compare drumming styles on particular songs, Pete Best and Ringo Starr have both recorded the following 33 songs. The Pete Best Band of the 1990s features two drummers, Pete and his brother, Roag. They play together on the recordings below.

The live recordings from the Star-Club in December 1962 were the subject of a court case in 1998. The Beatles gained the rights to the original tape and the TV-advertised 2CD set was withdrawn. The tapes had previously appeared as a double-album in 1977 and the sound had been considerably improved for CD with new technology. The exploitation of these tracks was the major issue, and I hope that, in time, the Beatles will realise that they are not their 'crummiest' performances and release them in a fully-annotated package.

Ask Me Why
* The Beatles with Pete Best (1962 recording on 'Pop Goes The Radio Vol 1', Italian CD)
* The Beatles with Ringo Starr (Parlophone single, 1962)

Back In The USSR
* The Beatles with Ringo Starr ('The Beatles' LP, 1968)
* The Beach Boys with Ringo Starr ('Fourth Of July' LP, 1986)
* The Pete Best Band ('Once A Beatle, Always...' CD, 1996)

Besame Mucho
* The Beatles with Pete Best (Decca audition tape, 1962)
* The Beatles with Pete Best (1962 Parlophone audition on 'Anthology 1' CD)
* The Beatles with Pete Best (1962 recording on 'Pop Goes The Radio Vol 1', Italian CD)
* The Beatles with Ringo Starr ('Live! At The Star-Club In Hamburg, Germany, 1962' LP)
* The Beatles with Ringo Starr (*Let It Be* film soundtrack, 1970)

Boys

* The Beatles with Ringo Starr ('Please Please Me' LP, 1963)
* The Beatles with Ringo Starr (1964 TV performance on 'Anthology 1' CD)
* The Beatles with Ringo Starr (1964 live recording on 'The Beatles At The Hollywood Bowl' LP)
* Ringo Starr and his All Starr Band ('Live From Montreux' CD, 1993)
* Pete Best (US Cameo single, 1966). Despite the label, vocalist is Wayne Bickerton.

Can't Buy Me Love

* The Beatles with Ringo Starr (1964 outtake on 'Anthology 1' CD)
* The Beatles with Ringo Starr (Parlophone single, 1964)
* The Beatles with Ringo Starr (1965 live recording on 'The Beatles At The Hollywood Bowl' LP)
* The Pete Best Band ('Once A Beatle, Always...' CD, 1996)

Crying Waiting Hoping

* The Beatles with Pete Best (Decca audition tape, 1962)
* The Beatles with Ringo Starr (1963 version on 'Live At The BBC' CD)

Dizzy Miss Lizzy

* The Beatles with Ringo Starr ('Help!' LP, 1965)
* The Beatles with Ringo Starr (1965 version on 'Live At The BBC' CD)
* The Beatles with Ringo Starr (1965 live recording on 'The Beatles At The Hollywood Bowl' LP)
* The Pete Best Band ('Back To The Beat' CD, 1995)

A Hard Day's Night

* The Beatles with Ringo Starr (1964 outtake on 'Anthology 1' CD)
* The Beatles with Ringo Starr (Parlophone single, 1964)
* The Beatles with Ringo Starr (1964 version on 'Live At The BBC' CD)
* The Beatles with Ringo Starr (1965 live recording on 'The Beatles At The Hollywood Bowl' LP)
* The Pete Best Band ('Once A Beatle, Always...' CD, 1996)

Hippy Hippy Shake

* The Beatles with Ringo Starr ('Live! At The Star-Club In Hamburg, Germany, 1962' LP)
* The Beatles with Ringo Starr (1963 version on 'Live At The BBC' CD)
* The Pete Best Band ('Back To The Beat' CD, 1995)

I'll Get You

* The Beatles with Ringo Starr (Parlophone B-side, 1963)
* The Pete Best Band (*Imagine...The Sixties* video 1990)

I'm Down

* The Beatles with Ringo Starr (1965 outtake on 'Anthology 2' CD)
* The Beatles with Ringo Starr (Parlophone B-side, 1965)
* The Pete Best Band ('Once A Beatle, Always...' CD, 1996)

I Saw Her Standing There

* The Beatles with Ringo Starr ('Please Please Me' LP, 1963)
* The Beatles with Ringo Starr (1963 version on 'Live At The BBC' CD)
* The Beatles with Ringo Starr (1963 live version on 'Anthology 1' CD)
* The Pete Best Band ('Back To The Beat' CD, 1995)

Johnny B Goode

* The Beatles with Ringo Starr (1964 version on 'Live At The BBC' CD)
* The Pete Best Band ('Back To The Beat' CD, 1995)

Kansas City

* The Beatles with Ringo Starr ('Live! At The Star-Club In Hamburg, Germany, 1962' LP)
* The Beatles with Ringo Starr (1964 outtake on 'Anthology 1' CD)
* The Beatles with Ringo Starr ('Beatles For Sale' LP, 1964)
* The Beatles with Ringo Starr (1963 version on 'Live At The BBC' CD)
* Pete Best (US Cameo single, 1966)

Long Tall Sally

* The Beatles with Ringo Starr ('Live! At The Star-Club In Hamburg, Germany, 1962' LP)
* The Beatles with Ringo Starr ('Long Tall Sally' EP, 1964)
* The Beatles with Ringo Starr (1963 version on 'Live At The BBC' CD)
* The Beatles with Ringo Starr (1964 TV performance on 'Anthology 1' CD)
* The Beatles with Ringo Starr (1964 live recording on 'The Beatles At The Hollywood Bowl' LP)
* The Merseybeats with Pete Best (*Merseybeat Magic* video 1989)
* The Pete Best Band ('Back To The Beat' CD, 1995)

Love Me Do

* The Beatles with Pete Best (1962 Parlophone audition tape on 'Anthology 1' CD)
* The Beatles with Ringo Starr, drums (Parlophone single, 1962)
* The Beatles with Andy White, drums and Ringo Starr, tambourine ('Please Please Me' LP, 1963)
* The Beatles with Ringo Starr, drums (1963 version on 'Live At The BBC' CD)
* The Pete Best Band ('Back To The Beat' CD, 1995)
* The Pete Best Band ('Once A Beatle, Always...' CD, 1996)
* Ringo Starr ('Vertical Man' CD, 1998)

Lucille

* The Beatles with Ringo Starr (1963 version on 'Live At The BBC' CD)
* The Pete Best Band ('Live At The Adelphi' CD, 1992)

Memphis Tennessee

* The Beatles with Pete Best (Decca audition tape, 1962)
* The Beatles with Pete Best (1962 recording on 'Pop Goes The Radio Vol 1', Italian CD)
* The Beatles with Ringo Starr (1963 version on 'Live At The BBC' CD)

Money (That's What I Want)

* The Beatles with Pete Best (Decca audition tape, 1962)
* The Beatles with Ringo Starr ('With The Beatles', LP, 1963)
* The Beatles with Ringo Starr (1963 live version on 'Anthology 1' CD)
* The Pete Best Band ('Back To The Beat' CD, 1995)

Mr Moonlight

* The Beatles with Ringo Starr ('Live! At The Star-Club In Hamburg, Germany, 1962' LP)
* The Beatles with Ringo Starr (1964 outtake on 'Anthology 1' CD)
* The Beatles with Ringo Starr ('Beatles For Sale LP, 1964)
* The Pete Best Band ('Once A Beatle, Always..' CD, 1996)

Please Mr Postman

* The Beatles with Pete Best (1962 recording on 'Pop Goes The Radio Vol 1', Italian CD)
* The Beatles with Ringo Starr ('With The Beatles' LP, 1963)

Please Please Me

* The Beatles with Ringo Starr (1962 outtake on 'Anthology 1' CD)
* The Beatles with Ringo Starr (Parlophone single, 1963)
* The Pete Best Band ('Once A Beatle, Always...' CD, 1996)

Revolution

* The Beatles with Ringo Starr (Apple B-side, 1968)
* The Pete Best Band ('Once A Beatle, Always...' CD, 1996)

Rip It Up

* The Beatles with Ringo Starr (1969 rehearsal on 'Anthology 3' CD)
* The Pete Best Band ('Live At The Adelphi' CD, 1992)

Rock And Roll Music

* The Beatles with Ringo Starr ('Beatles For Sale' LP, 1964)
* The Beatles with Ringo Starr (1964 version on 'Live At The BBC' CD)
* The Beatles with Ringo Starr (1966 live performance on 'Anthology 2' CD)
* Pete Best (US 'Best Of The Beatles' LP, 1965)
* The Pete Best Band ('Once A Beatle, Always...' CD, 1996)

Roll Over Beethoven

* The Beatles with Ringo Starr ('Live! At The Star-Club In Hamburg, Germany, 1962' LP)
* The Beatles with Ringo Starr ('With The Beatles' LP, 1963)
* The Beatles with Ringo Starr (1963 live version on 'Anthology 1' CD)
* The Beatles with Ringo Starr (1964 version on 'Live At The BBC' CD)
* The Beatles with Ringo Starr (1964 live recording on 'The Beatles At The Hollywood Bowl' LP)
* The Pete Best Band ('Back To The Beat' CD, 1995)

Slow Down

* The Beatles with Ringo Starr (1963 version on 'Live At The BBC' CD)
* The Beatles with Ringo Starr ('Long Tall Sally' EP, 1964)
* The Pete Best Band ('Live At The Adelphi' CD, 1992).

Some Other Guy

* The Beatles with Ringo Starr (1963 version on 'Live At The BBC' CD)
* Pest Best ('Best Of The Beatles' US LP, 1965)
* The Pete Best Band ('Once A Beatle, Always...' CD, 1996)

Sure To Fall

* The Beatles with Pete Best (Decca audition tape, 1962)
* The Beatles with Ringo Starr (1963 version on 'Live At The BBC' CD)
* Ringo Starr ('Stop And Smell The Roses' LP, 1981)

Ticket To Ride

* The Beatles with Ringo Starr (Parlophone single, 1965)
* The Beatles with Ringo Starr (1965 version on 'Live At The BBC' CD)
* The Beatles with Ringo Starr (1965 live TV performance on 'Anthology 2' CD)
* The Beatles with Ringo Starr (1965 live recording on 'The Beatles At The Hollywood Bowl' LP)
* The Pete Best Band ('Once A Beatle, Always...' CD, 1996)

Till There Was You

* The Beatles with Pete Best (Decca audition tape, 1962)
* The Beatles with Ringo Starr ('Live! At The Star-Club In Hamburg, Germany, 1962' LP)
* The Beatles with Ringo Starr, bongos ('With The Beatles' LP, 1963)
* The Beatles with Ringo Starr (1963 live version on 'Anthology 1' CD)
* The Beatles with Ringo Starr (1964 version on 'Live At The BBC' CD)

To Know Her Is To Love Her

* The Beatles with Pete Best (Decca audition tape, 1962)
* The Beatles with Ringo Starr ('Live! At The Star-Club In Hamburg, Germany, 1962' LP)
* The Beatles with Ringo Starr (1963 version on 'Live At The BBC' CD)

Twist And Shout

* The Beatles with Ringo Starr ('Please Please Me' LP, 1963)
* The Beatles with Ringo Starr (1963 live version on 'Anthology 1' CD)
* The Beatles with Ringo Starr (1965 live recording on 'The Beatles At The Hollywood Bowl' LP)
* The Pete Best Band and Merseycats ensemble (*Imagine...The Sixties* video 1990)
* The Pete Best Band ('Back To The Beat' CD, 1995)

I'VE JUST SEEN A FACE

To avoid repetition and to save you remembering everyone, this brief guide lists some of the personalities in *Drummed Out!*

Ron Appleby	1960s Southport promoter.
Neil Aspinall	Beatles' road manager and now manager of Apple Corps.
Tony Barrow	Beatles' press officer (1963-66).
Mona Best	Mother of Pete, Rory and Roag and owner of Casbah Club.
Roag Best	Pete Best's half-brother, born 1962. Also a drummer.
Wayne Bickerton	Member of Pete Best Four (1963-66). Producer of Rubettes.
Joe Brown	Singer-guitarist known for 'A Picture Of You'.
Howie Casey	Saxophonist, Seniors (1961-62), Kingsize Taylor and Dominoes (1963-64). Later with Wings.
John Cochrane	Drummer, Wump and his Werbles (1960-61), Midnighters (1962-63), Chuckles (1963-66).
Peter Cook	Lead guitarist, Kansas City Five (1962-63) and Earl Royce and Olympics (1963-65). Plays excellent blues guitar.
Lee Curtis	Lead singer with All Stars (1962-67). Owns guest house in Weston-Super-Mare and entertains residents.
Rod Davis	Quarry Men (1957-58).
Paddy Delaney	Doorman, Cavern Club.
Paul Du Noyer	Lennon biographer who edited *Q* magazine (1990-92).
Ray Ennis	Lead singer of Swinging Blue Jeans since 1957.
Clive Epstein	Younger brother of Brian Epstein. Died 1988.
Joe Flannery	Friend of Brian Epstein and manager of Beryl Marsden and his brother, Lee Curtis. Writing his autobiography.
Ritchie Galvin	Drummer, Earl Preston and TT's (1962-65). Now plays with Liverpool country bands.
Len Garry	Quarry Men (1957-58).
Johnny Gentle	Liverpool rock'n'roll-ballad singer managed by Larry Parnes.
Jim Gretty	Guitar tutor who sold instruments at Frank Hessy's.
Eric Griffiths	Quarry Men (1957-58).
Johnny Guitar	Rhythm guitarist with Rory Storm and Hurricanes. Despite industrial injury to his shoulder, still plays Merseyside clubs.
Johnny Gustafson	Cass and Cassnovas (1959-60), Big Three (1961-63), Merseybeats (1964).
Colin Hanton	Quarry Men (1957-59).
Bill Harry	Editor *Mersey Beat* newspaper and Beatles chronicler since.
Johnny Hutchinson	Cass and Cassanovas (1959-60), Big Three (1961-64).
Bert Kaempfert	MOR German orchestra leader and record producer.
Brian Kelly	Impresario who ran dances in the northern suburbs of Liverpool. He ran Alpha Sound in Crosby.
Astrid Kirchherr	German photographer and Stu Sutcliffe's girlfriend.
Ted Knibbs	Manager of Billy J Kramer. Sold contract to Eppy for £50.
Billy J Kramer	With Coasters (1960-63) and Dakotas (1963-67, 1997-present).
Sam Leach	Entertaining Merseybeat impresario. Calls himself The Man That Merseybeat Forgot and thinks he's being ironic.